Mechanical Measurements

(THIRD EDITION)

R.S. Sirohi

Professor, Engineering Design Center

AND

H.C. Radha Krishna

Professor, Mechanical Engineering Department
Indian Institute of Technology, Madras
India

JOHN WILEY & SONS

NEW YORK CHICHESTER BRISBANE TORONTO SINGAPORE

First Published in 1991 by
WILEY EASTERN LIMITED
4835/24 Ansari Road, Daryaganj
New Delhi 110 002, India

Distributors:

Australia and New Zealand:
JACARANDA WILEY LIMITED
P.O. Box 1226, Milton Old 4064, Australia

Canada:
JOHN WILEY & SONS CANADA LIMITED
22 Worcester Road, Rexdale, Ontario, Canada

Europe and Africa:
JOHN WILEY & SONS LIMITED
Baffins Lane, Chichester, West Sussex, England

South East Asia:
JOHN WILEY & SONS (PTE) LIMITED
05-04, Block B, Union Industrial Building
37 Jalan Pemimpin, Singapore 2057

Africa and South Asia:
WILEY EASTERN LIMITED
4835/24 Ansari Road, Daryaganj
New Delhi 110 002, India

North and South America and rest of the World:
JOHN WILEY & SONS, INC.
605 Third Avenue, New York, NY 10158, USA

Library of Congress Cataloging-in-Publication Data

ISBN 0-470-21953-X   John Wiley & Sons, Inc.
ISBN 81-224-0383-2   Wiley Eastern Limited

Printed in India at A.P. Offset, Delhi 110 032.

Foreword

Experimental measurements, methods and techniques are becoming increasingly important in engineering education. The ability to perform experimentation and take measurements with acceptable precision deserve more emphasis. In recent years laboratory programmes have been modernized, sophisticated electronic instrumentation are incorporated into the programme and newer techniques have been developed. In as much as a well designed laboratory programme is essential in the undergraduate engineering education providing the students with laboratory manuals, guides and text books is equally important. It is in this context that this book makes its timely appearance, which, I am sure, will be gladly received by all. My senior colleagues, Dr. R.S. Sirohi and Dr. H.C. Radha Krishna, are to be congratulated for their efforts in bringing out this book to meet the needs of the students and teachers in comprehending the various aspects of mechanical measurements.

Indian Institute of Technology
Madras

DR. M.C. GUPTA
Professor of Mechanical Engineering
and
Coordinator Mechanical Engineering
Education Development Center

Preface to the Third Edition

The third edition of the book contains an additional chapter on 'Fibre-Optics in Measurements'. Fibre-optic sensors are finding numerous applications in industry to monitor various variables. It was therefore felt that the information on FO sensors should be made available to the readers. The chapter deals with the basic physics of light propagation in fibres, fibre types and a number of FO sensors and their application areas. It is hoped that this chapter would enthuse the readers to know more about the various possibilities of FO sensors in measurement.

Madras
March 1991

R.S. SIROHI
H.C. RADHAKRISHNA

Preface to the Second Edition

Considering the suggestions from many readers and colleagues, the text has been revised. Some of the sections are rearranged and rewritten for a greater clarity. A new chapter that includes some of the recent methods of measurements by Optical Techniques has been added. This book has taken the shape it has by the large amount of available information on this subject in literature. A bibliography is given at the end of the book. In view of the exceptionally good method of presentation in some of these publications, the style of presentation of the relevant portions has been followed here also. We gratefully acknowledge the Authors and the Institutions responsible for these publications. Many of our colleagues have been helpful during the course of rewriting the book. We acknowledge their help.

R.S. SIROHI
H.C. RADHA KRISHNA

Preface to the First Edition

Engineers are involved in the design of systems, systems involving many components. The effectiveness of the system is verified by Measurements. Measurement is an Art, a Science and a Technique by itself. There is nothing which can be effected without Measurement. An attempt is made here to deal with a part of the knowledge commonly identified as Mechanical Measurements. The stress is 1 laid on the understanding of the physics of the Measurement Techniques.

Every measurement is in error. For the measurement to be meaningful, the nature and magnitude of the error should be known. Therefore the book begins with error analysis and the application of statistical principles to measurements.

The methods of measuring various mechanical quantities are discussed subsequently, covering both the basic and derived quantities. Effort has been made to present the subject in S.I. units. The coverage in the book is such that it may be successfully used as a text both for graduate and post-graduate classes and as a constant reference to a researcher.

If the student realises the importance and understands the methods of measurement required by him, we will be pleased and honoured. The material contained in this book is extracted freely from books, journals and pamphlets. We gratefully acknowledge all of them. Professor M C Gupta, Professor of Mechanical Engineering and Coordinator, Curriculum Development Center, Indian Institute of Technology, Madras, spontaneously responded to our request to publish this first in the form of a monograph. We are thankful to him for his encouragement and support. Many of our colleagues have been of immense help in this effort of ours. We acknowledge all of them.

R.S. SIROHI
H.C. RADHA KRISHNA

Contents

Basic Concepts of Measurements

1.1 What is measurement?

In order to compare or determine the value of a physical variable, some kind of measurement is to be carried out. Fundamentally, measurement of a quantity is the act or the result of a quantitative comparison between a predefined standard and an unknown magnitude. If the result is to be meaningful, the act of measurement must satisfy the following requirements:

    (1)   the standard must be accurately known and internationally accepted, and

    (2)   the apparatus and the experimental procedure adopted for comparison must be provable.

In order to be able to consistently compare, standards of length, mass, time, temperature and electrical* quantities have been established. These standards are internationally accepted, and some of them are preserved under controlled environmental conditions. The procedures employed for comparison are well documented.

The standard of length is the standard metre defined as a length between two very fine lines engraved on a platinum-iridium bar of X cross-section maintained and measured under very accurately controlled conditions. The General Conference of Weights and Measures defined the standard metre in terms of the wavelength of orange-red light of Kr^{86} lamp. The standard metre is eqnivalent to 1,650,763.73 wavelengths of Kr^{86} orange-red light. Presently the metre is being redefined in terms of the wavelength of He-Ne laser. The kilogram is defined in terms of platinum-iridium mass. Both the standard metre and kilogram mass are maintained at the International Bureau of Weights and Measures in Sevres, France. The Cesium clock has been accepted as the standard of time measurement.

An absolute temperature scale was conceived by Lord Kelvin in 1854 on the basis of the second law of thermodynamics. The international temperature scale of 1948 furnishes an experimental basis of a

---

*The standard electrical units are derivable from the mechanical units of mass, length and time.

temperature scale which approximates as closely as possible with the absolute thermodynamic scale.*

1.2   Methods of measurements

The two basic methods of measurement are: (a) a direct comparison with the primary or the secondary standard and (b) an indirect comparison with a standard, through a calibrated system. Any one of these two methods can be employed, depending on the requirements. But to restrict the frequent use of primary standards, secondary standards are generally used for direct comparison or calibration of a system.

1.3   Calibration

A measuring instrument is to be checked for accuracy at frequent intervals with a known standard, and any discrepancy between the measured value and known standard to be set right through calibration. Calibration procedures thus involve a comparison of a particular instrument with either (i) a primary standard, or (ii) a secondary standard, or (iii) a known input source. The secondary standard employed for calibration must possess an accuracy at least ten times higher than that of the instrument to be calibrated. The calibration procedures for various instruments will be described at the appropriate places.

1.4   Why make measurements?

One of the basic functions of all engineering branches is design—the design of systems consisting of several elements that are expected to function in a (desired) fashion. Measurement is required to test if the elements that constitute the system function as per the design expectation and finally to evaluate the functioning of the system itself.

Further, in basic sciences, fundamental principles or derived phenomena are studied in great detail to confirm the validity of certain postulates. Sometimes in anticipation of discovering some strange behaviour of nature or proving the validity of a theory, a sophisticated instrumentation is carried out. The analysis is carried out on numerical data which arises due to the act of measurement.

To propound a statistical law, measurements are made on a number of systems.

The measuring instrument is an essential component of a control system in which the measured variable is constantly compared with the set point.

Hence measurements are essential for evaluating the performance of a

*Appendix contains information about the range and accuracy of standards of relevant quantities maintained at National Physical Laboratory, New Delhi.

system, or studying its response to a particular input function, or studying some basic law of nature etc.

1.5 Generalised description of a measurement system

The act of measurement is accomplished with a measuring instrument, an assemblage of physical facilities. An instrument is designed to perform a certain task and its description is alway possible in terms of its physical elements. One of the major shortcomings of this approach is that it demands a separate description of each instrument. Each of the physical elements that constitute an instrument can be identified with a functional element. It is possible that a physical element may perform many functions. The description of an instrument in terms of its functional elements falls under the generalised approach. There could be various schemes to describe an instrument in terms of its functional elements, but the one presented below is found to be quite adequate and simple.

A block diagram representation of a generalised measurement instrumentation is shown in Fig. 1.1. An instrument to measure a quantity (2) of a medium (1) may comprise of:

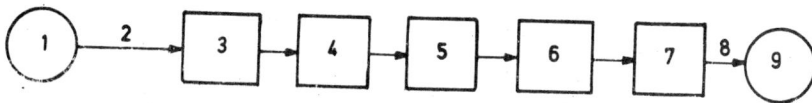

Fig. 1.1 Block diagram of a generalised measurement instrumentation
1. Measured medium, 2. Measured quantity, 3. Primary sensing element, 4. Variable conversion element, 5. Variable manipulation element, 6. Data transmission element, 7. Data presentation element, 8. Presented data, and 9. Observer

(a) PRIMARY SENSING ELEMENT (3)

This is an element which first receives energy from the medium to be measured and produces a proportional output. The output signal of the primary sensing element is a physical variable such as displacement or voltage. The primary sensing element, therefore, is a primary transducer (sensor) that converts (transduces) one physical variable or effect into another. A primary transducer may be followed by an intermediate transducer, if a second transduction is desired. For the measurement to be faithful, the sensor should be so designed as to extract a very small amount of energy from the medium. In other words, the medium should not be disturbed appreciably when the sensing element is inserted. As an example, consider the measurement of temperature of a body. The heat capacity of the thermometer, if large, will lower the temperature of the body and hence the measured temperature will be considerably smaller than its actual temperature. Table 1.1 gives a number of variables transduced to other physical variables, and the systems based on them.

TABLE 1.1   Transduction operation of a few physical variables

Transduction		Systems based on them
from	to	
Displacement	Displacement	contacting spindle, pin, light pointer
	Resistance change	strain gauges
	Capacitance change	capacitance strain gauges, torque meter
	Inductance change	inductance strain gauges, rotameters, differential transformers
	Voltage	piezo-electric probes
	irradiance change	interferometers
Force	Displacement	proving rings, springs
Pressure	Displacement	Bourdon tubes, bellows, diaphragms
Flow	Pressure	orifice plate, nozzles, venturi, pitot tube
	Displacement	piston type flowmeters
	Rotation	turbin flowmeters
	Voltage	magnetic flowmeters
	Frequency	Laser Doppler anemometer, Ultrasonic flowmeter
	Temperature change	hot-wire anemometer
Temperature	Displacement	differential expansion temperature sensors, bimetallic sensors
	Electric current	thermocouples and thermopiles
	Resistance change	resistance thermometers
	Pressure	pressure thermometers
	Temperature	IR pyrometers

(b)   VARIABLE-CONVERSION ELEMENT (4)

The output signal of the primary sensing element may require to be converted to a more suitable variable while preserving its information contents. This function is performed by the variable conversion element and it may be considered as an intermediate transducer. An example of this is the conversion of displacement to voltage in a pressure transducer.

(c)   VARIABLE-MANIPULATION ELEMENT (5)

This element is an intermediate stage of a measuring system. It modifies the direct signal by amplification, filtering etc. so that a desired output is produced. The physical nature of the variable remains unchanged during this stage. An example of this is an ac amplifier tuned to the frequency of chopper in some spectro-photometers.

(d) Data-transmission Element (6)

When the functional elements of a measuring system are spatially separated, it becomes necessary to transmit signals from one element to another. This function is performed by the data-transmission element. It is an essential functional element where remote controlled operation is desired.

(e) Data-presentation Element (7)

Usually information about the quantity being measured is to be communicated to a human observer for monitoring, control or analysis purposes. This is, therefore, to be presented in a form recognisable by some human senses. If the information is to be presented to the computer, it can be done in the form of a binary scale on the punched tape or cards, or the measuring system may be suitably interfaced with the computer. An element that performs this 'translation' function is called a data-presentation element. The observer (9) receives the data (8) in a form he can interpret easily.

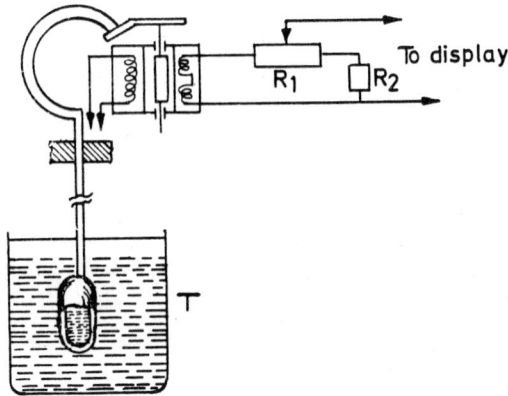

(a)

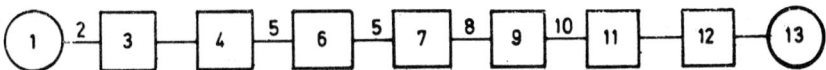

(b)

Fig. 1.2 (a) Pressure thermometer and (b) Its functional description
1. Measured medium, 2. Temperature-quantity to be measured, 3. Primary sensing element, 4, 7, 9. Variable conversion element, 5. Pressure, 6. Data Transmission element, 8. Motion, 10. Voltage, 11. Variable manipulation element, 12. Data presentation element, and 13. Observer.

Bulb	3, 4
Tubing	6
Bourdon tube	7
Differential transformer	9
Amplifier	11
Display unit	12

An example given in Fig. 1.2 may illustrate the working of this representation. A pressure type thermometer [Fig. 1.2 (a)] is used to measure the temperature of a fluid. The thermometer works on the principle of differential expansion of liquid which in turn imparts pressure to the Bourdon tube. The displacement of the free end of the Bourdon tube is magnified by a differential transformer transducer and its output is given to a display device. Fig. 1.2(b) is a block diagram representation of the measurement act. It may be emphasized that the same physical element can perform more than one function and the functions need not be performed in the sequence of Fig 1.1.

1.6 Operational description of a measurement system

An overall description of the performance of an instrument can be given in terms of its operational transfer function. An instrument performs an operation on an input quantity to provide an output. A block diagram of this is illustrated in Fig. 1.3.

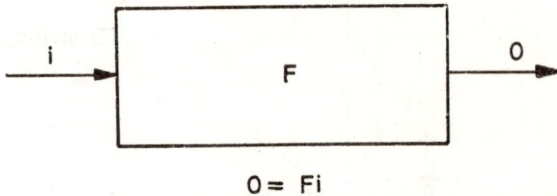

i → | F | → O

$$O = Fi$$

Fig. 1.3 Input-Output relation of a measurement system

The input-output relationship is characterised by the operation F such that

$$o = F_i$$

where o is the output for the input i. The symbol F refers to an operation on the input to the system and the description of the system is contained in this operation.

The input to the instrument may be any one or the combination of the following types:

(i) *Desired input:* This is an input which the instrument is specially designed to measure.

(ii) *Interfering input:* This represents an input for which the instrument is unintentionally responsive.

(iii) *Modifying input:* This modifies the input-output relationship for both the desired and interfering inputs.

It is most likely that interfering and modifying inputs are present along with the desired input. The integrated influence of these inputs on the instrument's output is shown in Fig. 1.4. F_I, F_D and F_M refer to the system's operational functions for interfering, desired and modifying inputs respectively.

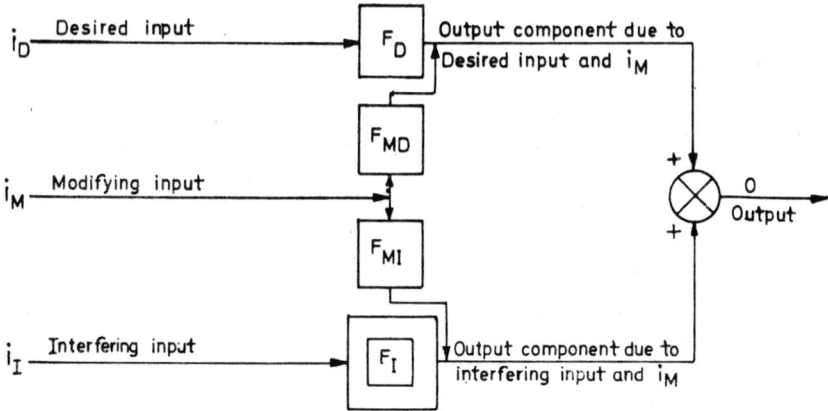

Fig. 1.4   Generalised input-output configuration

The physical meaning of these inputs can be grasped by considering an example from the measurement field. Consider an electrical strain gauge set up (Fig. 1.5) used for the measurement of strain induced due to pressure in a pipe. The strain ϵ is measured in terms of the resistance change ΔR of the strain gauge. It is given by

$$\epsilon = \frac{1}{G_F}\left(\frac{\Delta R}{R}\right),$$

where R is the resistance of the gauge and G_F the gauge factor. The resistance change (ΔR) is measured with Wheatstone bridge arrangement. In this particular example strain ϵ is the desired input and the bridge output e_0 is the output which is proportional to ΔR.

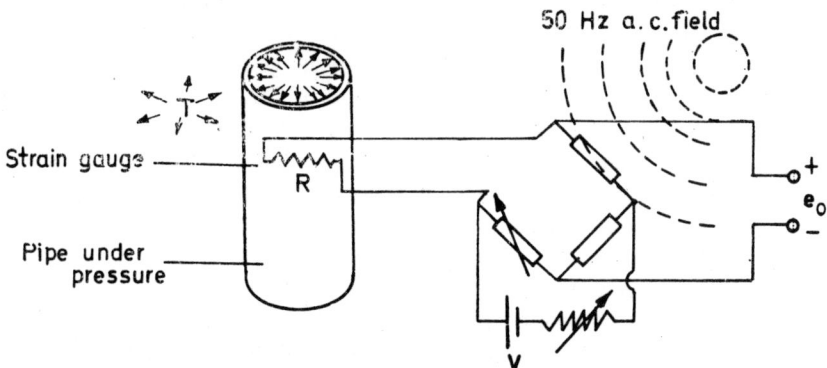

Fig. 1.5   Interfering and modifying inputs for strain gauge

An interfering input to this instrument is 50 Hz stray field which can induce voltage that may appear as output although the strain is zero. Another interfering input is temperature. A change in the ambient temperature results in the resistance change of the strain gauge and consequently an output is developed even in the absence of strain. The gauge factor G_F is temperature dependant. Therefore both the desired and interfering inputs are influenced by the changes in the ambient temperature. The temperature is a modifying input as well in this example.

Often it is desired to eliminate or reduce the influence of undesired inputs on the output of an instrument. There are a number of methods to accomplish this. A few of them are described below:

(i)   *Method of inherent insensitivity*

The design of an instrument, particularly the primary sensing element, should be such that it is sensitive only to desired input. In the example of Fig. 1.5, the strain-gauge may be made of a material which has a very low temperature coefficient of resistance, and temperature independent gauge factor, while retaining the sensitivity to strain. The instrument may be shielded from the stray field.

(ii)   *Compensation method*

This method consists of intentionally introducing into the instrument interfering and/or modifying inputs that tend to cancel the undesired effects of the unavoidable spurious inputs. The influence of temperature, in the example of Fig. 1.5, is removed by inserting an identical strain gauge into another arm of the Wheatstone bridge and exposing it to the identical environmental conditions. There are many instrumentation areas which employ the compensation techniques viz., resistance and thermocouple thermometry, photomultiplier circuitary, etc.

(iii)   *Output correction by analysis*

This method requires the knowledge of the mathematical relation between the spurious input and its output. If the temperature coefficient of resistance, temperature sensitivity of the gauge factor and the temperature of the gauge are known in the example of Fig. 1.5 then a correction to the output can be calculated and applied. Similarly a correction due to the exposed portion of the mercury in a glass thermometer can be made, if the thermometer is used in a manner other than recommended by the calibration procedure.

(iv)   *Signal filtering*

This method is based on the possibility of introducing certain elements (filters) into the instrument which block the spurious inputs so that their influence on the output is either eliminated or minimised. The filters can be inserted either in the path of the spurious inputs or output path.

Processing of data may also involve this step. In the example of strain gauge, the spurious input due to 50 Hz stray field can be removed by electrically shielding the instrument. The instruments sensitive to vibrations can be mounted on damping devices, the effect of tilt can be removed by gyroscopic mounting, the effect of d.c. (ambient light) in pyrometry or spectrophotometry can be removed by chopping the incoming beam etc.

1.7 Signal types

(i) *Analog signals:* The output of an instrument can be either analog or digital. The analog signals vary in a continuous fashion. They can take infinite values in a given range. The devices that produce such signals are called analog devices. Most of the devices or instruments which are used for measurement and control are of the analog type.

(ii) *Digital signals:* The digital signals vary in discrete steps and thus can take up only a finite number of different values in a given range. In a digital instrument the signals are automatically digitised and the result is presented as a digital read-out. Due to the application of digital computers for date handling, reduction and analysis, and in automatic control, the importance of digital instrumentation is growing very fast. The data to a digital computer is to be provided in the digital form, while most of the measuring instruments provide output in an analog form. Therefore conversion from analog to digital form is required. This is performed by an analog to a digital converter. Similarly, to convert a digital signal to an analog signal, a digital-to-analog converter is needed. Both A/D and D/A converters serve as 'translators' that enable the digital computer to communicate with the outside world that is largely of analog nature. All microprocessor based instrumentation is essentially of digital nature.

1.8 Modes of operation

(i) *Deflection mode:* The measuring instruments are used either in deflection or null mode. In deflection mode, the quantity to be measured produces some physical effects on a part that causes a similar but opposing effect on some other part of the instrument. At balance, the opposing effect is equal to that produced by the measured quantity. An example of such an instrument is a moving coil galvanometer, where the force experienced by the coil carrying a current, is counterbalanced by the torsional force of the suspension filament. At balance, the deflection of the light beam or the pointer on a scale gives the magnitude of the current. All deflection type instruments are initially calibrated, and the value of the measured quantity is read off from the deflection position on the scale.

(ii) *Null mode:* In contrast to the deflection type instruments, a null type instrument maintains balance at one point, between the effect generated by the measured quantity and the suitable opposing effect applied to it. Knowledge of the quantity that produces opposing effect provides the value of the quantity to be measured. For reasons of accuracy null type instruments are preferred over the deflection type instruments.

Exercises

1. Define the word transducer. What do you understand by active and passive transducers? Give examples and explain their relative merits and demerits.
2. The data to the observer can be presented either in analog or digital form. Mention some methods for analog to digital conversion.
3. Null method of measurement is often preferred to the deflection method. Comment.
4. With reference to Fig. 1.5, draw a Wheatstone bridge circuit diagram where temperature compensation is achieved by the use of a dummy strain gauge. Provide necessary mathematical support for your choice of the placement of dummy strain gauge.
5. The pressure thermometer of Fig. 1.2 (a) is used to measure the temperature of a remotely located hot body. The variation of ambient temperature will influence the temperature of fluid in the data transmission element, i.e. tubing and hence the measured temperature. How do you compensate for the influence of ambient temperature variation?
6. Discuss the potentiometric method of voltage measurement. Compare it with the direct method using a volt meter.

2

Performance Characteristics of Measuring Instruments

2.1 Introduction

The response of an instrument to a particular input is the guiding factor to decide its choice out of a number of available instruments or to venture on a new design. The input to the instrument could be either constant or rapidly varying with time. Therefore the performance of an instrument is discussed usually under the following two heads:

    (1) Static performance characteristics, and
    (2) Dynamic performance characteristics.

2.2 Static performance characteristics

An instrument may be used to measure quantities which are either constant or vary very slowly with time. The choice of the instrument for such a case is made by its static calibration i.e., by its input-output relationship.

STATIC CALIBRATION

Static calibration refers to a procedure where an input, either constant or slowly time varying, is applied to an instrument and corresponding output measured, while all other inputs (desired, interfering, modifying) are kept constant at some value. Instruments are so constructed that the signal conversion they perform have the property of irreversibility or directionality. This implies that a change in an input quantity will cause a corresponding change in the output. The functional relationship between the output quantity q_0 and the input quantity q_i is referred to as static calibration valid under the stated constant conditions of all other inputs. The static calibration may be expressed analytically ($q_0 = f(q_i)$), or graphically, or in a tabular form. A graphical representation between q_0 and q_i is the calibration curve applicable under the stated constant conditions for all other inputs. The static performance characteristics are obtained from the calibration curves. It may, however, be emphasised that a calibration standard must be at least ten times more accurate than the instrument to be calibrated. Some of the static performance characteristics are

(i)   *Linearity*

If the relationship between the output and input can be expressed by an equation of the form

$$q_0 = a + kq_i,$$

where a and k are constants, the instrument is said to possess linearity. Linearity, in practice, is never completely achieved, and the deviations from the ideal are termed as linearity tolerances. In commercial instruments, the maximum departure from linearity is often specified. Independent linearity and the proportional linearity are the two forms of specifying linearity. They are illustrated in Fig.2.1(a) and 2.1(b). As an example, 3% independent linearity means that the output will remain within values set by two parallel lines spaced ±3% of the full scale output from the idealised line (Fig.2.1 (a)). Similarly, 3% proportional linearity is illustrated in Fig.2.1(b). The ideal value is never more than ±3% away from the recorded value regrdless of the magnitude of the input. If the input-output relation is not linear for an instrument, it may still be approximated to a linear form when it is used over a very restricted range. It may, however, be noted that an instrument which does not possess linearity can still be highly accurate. In some instruments like ring-balance meter where the input-outpnt relation is essentially non-linear, linearisation is achieved mechanically or electrically at least over the limited range.

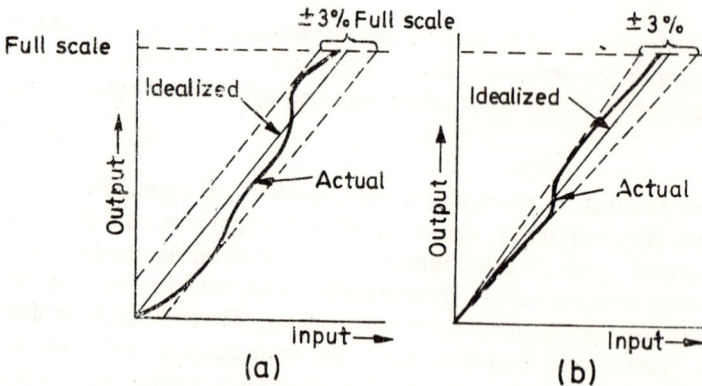

Fig. 2.1   Specifying linearity   (a) Independent   (b) Proportional

(ii)   *Static sensitivity*

The static sensitivity is defined as the slope of the calibration curve i.e.

$$\text{Sensitivity} = \lim_{\Delta \to 0} \frac{\Delta q_0}{\Delta q_i} = \frac{dq_0}{dq_i}.$$

If the input-output relation is linear, the sensitivity is constant for all values of input. The sensitivity of an instrument having a non-linear static characteristics depends on the value of the input quantity and should be specified as

$$\text{Sensitivity} = \lim_{\Delta \to 0} \frac{\Delta q_0}{\Delta q_i}\bigg|_{q_i}$$

These definitions are graphically illustrated in Fig. 2.2(a) and 2.2(b). The calibration curves are plotted using least square method from the experimental data points.

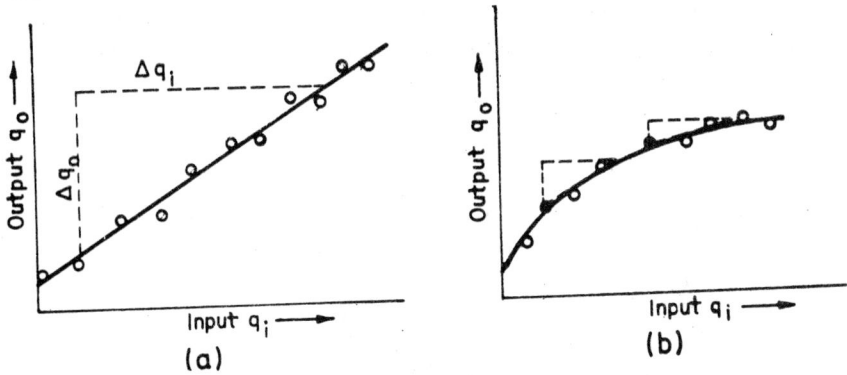

Fig. 2.2 Definition of static sensitivity

While the instrument's sensitivity to its desired input is of primary importance, its sensitivity to interfering, and modifying inputs may also be of considerable interest. As an example. consider the case of the strain gauge already discussed in chapter 1. The temperature is an interfering input, and causes the resistance of the gauge to vary and thus would drift the output value even when the strain is zero. This is called the zero drift. Further, the temperature is also a modifying input which

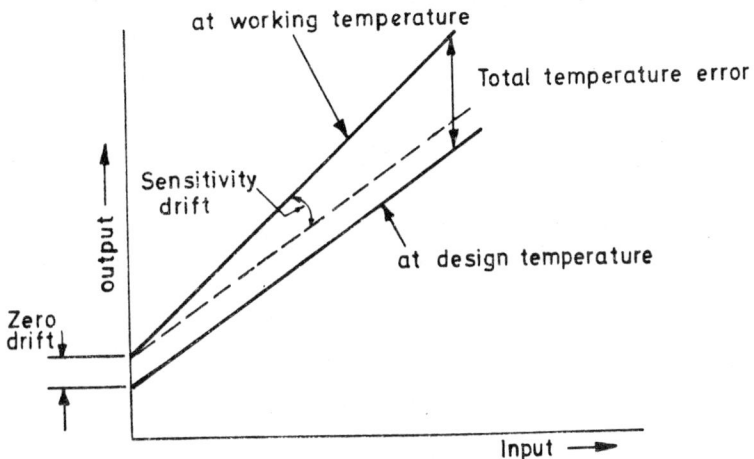

Fig. 2.3 Definitions of zero drift and sensitivity drift—influence of temperature on the strain gauge

changes the sensitivity factor. This effect is called the sensitivity drift or scale factor drift. Fig. 2.3 illustrates these definitions clearly. The total error due to temperature on the strain measurement has also been shown in the Fig. 2·3.

(iii) *Repeatability*

If an instrument is used to measure same or an identical input many times and at different time intervals, the output is not the same but shows a scatter. This scatter or deviation from the ideal static characteristics, in absolute units or a fraction of the full scale, is called repeatability error and is illustrated in Fig. 2.4.

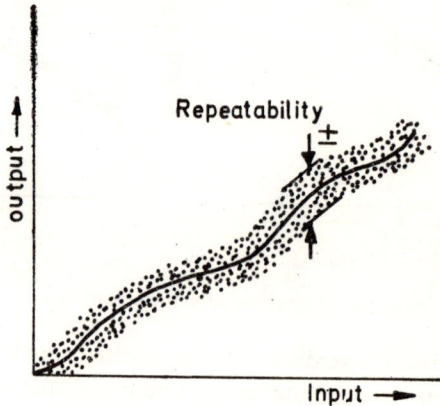

Fig. 2.4 Repeatability error

(iv) *Hysteresis-threshold-resolution*

While testing an instruement for repeatability, it is often seen that input-output graphs do not coincide for continuously ascending and then descending values of the input. This non-coincidence of input-output graphs for increasing and decreasing inputs arises due to the phenomenon of hysteresis (Fig. 2.5). Some causes for hysteresis effect in an instrument are internal friction, sliding or external friction, free play or

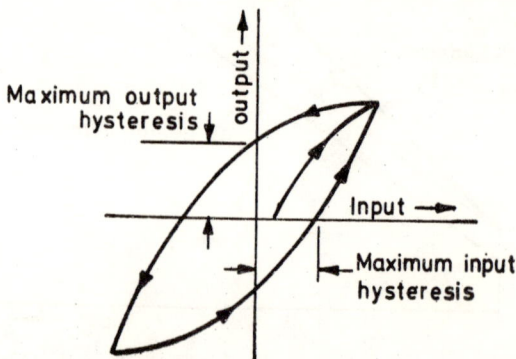

Fig. 2.5 Hysteresis effect

looseness of a mechanism, etc. Hysteresis effects are best eliminated by taking readings corresponding to ascending and descending values of the input and then taking their arithmatic average.

Threshold and resolution are the other two characteristics of a measuring instrument. In order to understand them consider an instrument to which an input is applied which is gradually increased from zero. It will be observed that there is a certain minimum value of the input below which no output change can be detected. This minimum value is taken as threshold input of the instrument. However, a statement like detectable output change is very vague. Therefore, to provide a meaningful definition of the term threshold and to improve reproducibility of the measurement of threshold certain minimum output change is prescribed and the corresponding input is called threshold. Similarly, if the instrument is being used for measurement, there is a minimum change in the input for which certain detectable change in the output is observed. This incremental change in input is referred as resolution. Threshold is measured when the input is varied from zero while the resolution is measured when the input is varied from any arbitrary non-zero value. Therefore the threshold is the smallest measurable input while the resolution is the smallest measurable input change. Both the threshold and the resolution can be given either as absolute values or as percentage of full scale deflection.

(v) *Readability and span*

The readability depends both on the instruments and observer and often is not stated. The span refers to the range of the instrument.

2.3 Dynamic performance characteristics

Some properties of measuring instruments when a static or slowly time varying input is applied to them have been discussed. However when a time varying input is applied to the instrument, the output may be distorted not only by static error but also by dynamic error. An instrument designer should know the response of the instrument to the dynamic input. In fact often the instrument is used under dyanmic conditions. The difference between the static output and the dynamic output corresponding to an unknown input at a given instant of time is defined as the dynamic error. If the input is $q_i(t)$ and output $q_0(t)$, then the absolute dynamic error at any time in the absence of static errors is given by

$$\Delta q_e(t) = q_{0s}(t) - q_0(t),$$

where $q_{0s}(t)$ is the output of an ideal instrument and is given by $q_{0s}(t) = kq_i(t)$. The dynamic error depends on the input and will be discussed subsequently along with the instruments types.

Mathematically, the dynamical behaviour of an instrument is describad by a linear differential equation with constant coefficients. A general relation between any particular input and the corresponding output, under suitable simplifying assumptions, can be expressed by a differential equation of the form

$$(a_n D^n + a_{n-1} D^{n-1} + \ldots + a_1 D + a_0)q_0 = (b_m D^m + b_{m-1} D^{m-1} + \ldots + b_1 D + b_0)q_i,$$

where a's and b's are physical constants, $D (= d/dt)$ is the differential operator, and both input q_i and output q_0 are functions of time. The task is to find out q_0 in terms of system parameters and the input q_i. Solution of the above differential equation can be written as

$$q_0 = q_{ocf} + q_{opi}$$

where q_{ocf} is complimentary function part of the solution, and q_{opi} is particular integral part of the solution. The complimentary function part is the solution when no input is applied. This gives the natural behaviour of the system. The particular integral part is the solution due to the impressed input.

The methods of solving a general linear differential equation will not be discussed here. Instead solutions of differential equations pertaining to particular cases as and when they arise will be discussed. It is, however, worthwhile here to introduce the concept of transfer function and frequency response.

Operational transfer function

The operational transfer function relating the output q_0 to the input q_i is defined by the equation

$$\frac{q_0}{q_i}(D) = \frac{b_m D^m + b_{m-1} D^{m-1} + \ldots + b_1 D + b_0}{a_n D^n + a_{n-1} D^{n-1} + \ldots + a_1 D + a_0}$$

The representation $\frac{q_0}{q_i}(D)$ is a very general relation. The operational transfer function is a very useful concept in the analysis, design and application of an instrument. The merits of operational transfer function approach are:

(a) The dynamic characteristics of the system can be represented by means of block diagrams

(b) It is very helpful to determine the overall characteristics of the system in terms of the transfer functions of its components. If the loading effects are negligible, the overall transfer function is the product of the transfer functions of the individual components.

Sinusoidal transfer function and frequency response

If an input of the form

$$q_i = A_i \sin \omega t$$

is applied to an instrument it will be seen, when all the transient effects die out, that the output q_0 will be a sinusoidal wave of the same frequency ω as that of the input. Usually both the amplitude and phase of the output will be different from that of the input. Since the frequency is the same, the input-output relation is completely described in terms of the amplitude ratio and phase shift. Phase shift is the difference between the phases of input and output. Both the amplitude ratio and phase shift vary, in general, with the frequency ω. Thus the frequency response of a linear system consists of curves of amplitude ratio and phase shift as a function of frequency. The sinusoidal transfer function of an instrument is defined as:

$$\frac{q_0}{q_i}(i\omega) = \frac{b_m(i\omega)^m + b_{m-1}(i\omega)^{m-1} + \ldots + b_1(i\omega) + b_0}{a_n(i\omega)^n + a_{n-1}(i\omega)^{n-1} + \ldots + a_1(i\omega) + a_0}$$

The quantity $\frac{q_0}{q_i}(i\omega)$ is a complex quantity. Its magnitude $\left|\frac{q_0}{q_i}(i\omega)\right|$ gives the amplitude ratio while its argument is the phase. The relationship between this complex quantity and the frequency constitutes the frequency response.

2.4 Input types

Although a measured quantity, in general, will not be a simple function of time, much can be learnt about an insturment by observing its responce to some elementary inputs. Four types of elementary inputs are described here:

Step input

It is mathematically represented as

$$q_i = 0 \text{ for } t < 0$$
$$= q_{is} \text{ for } t > 0$$

where q_{is} is a constant value. The input jumps to a constant value in an infinitesimally small time. This is illustrated in Fig. 2.6 (a).

Ramp input

The mathematical representation of this is

$$q_i = 0 \text{ for } t < 0,$$
$$= q_{is} t \text{ for } t \geqslant 0,$$

where q_{is} is the slope of input vs time relation. The input linearly increases with time as shown in Fig. 2.6 (b).

Sinusoidal input

The sinusoidal input is represented by the equation

$$q_i = A_i \sin(\omega t + \delta).$$

where A_i is the amplitude, ω the frequency and δ the phase. This is one of the very important elementary inputs. Any input can be analysed in terms of its sinusoidal components using Fourier analysis, and response of an instrument to sine inputs of different frequencies generates its frequency response. The sinusoidal input for $\delta = 0$ is graphically shown in Fig. 2.6(c).

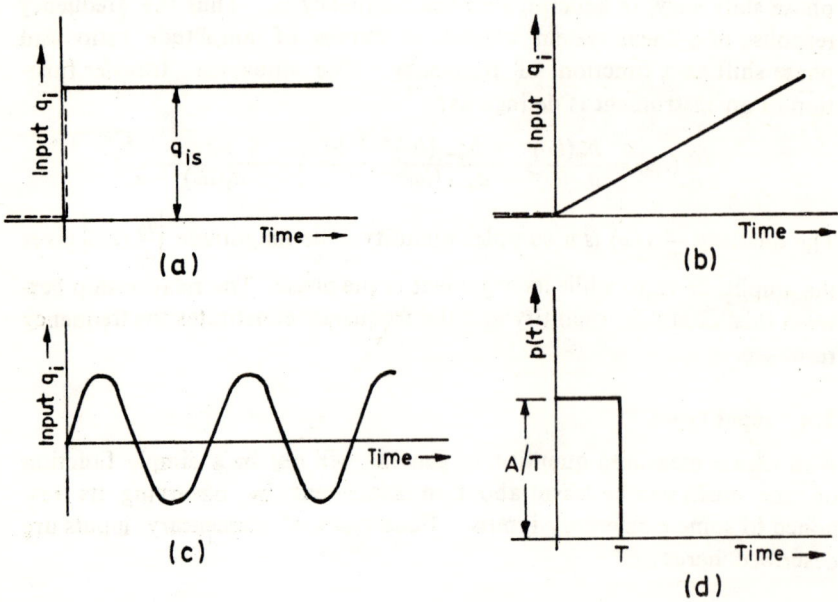

Fig. 2.6   Elementary inputs (a) Step, (b) Ramp, (c) Sinusoidal, and (d) Impulse

Impulse input

An impulse functions of strength A is defined by $\triangleq \lim_{T \to 0} p(t)$ and is illustrated in Fig. 2.6(d). It has the following properties:

(a)   The duration of the pulse is infinitesimally short $(T \to 0)$.

(b)   The peak of the pulse is infinitely high $(A/T \to \infty)$.

(c)   The area of the pulse is finite and is equal to the strength A of the pulse.

If the area of the pulse is unity, it is called unit impulse function $u_1(t)$. Thus $p(t) = Au_1(t)$ [Fig. 2.6(d)]. The impulse function is very useful in studying the frequency response of mechanical instruments as it possesses all the frequency components.

2.5   Instruments types

Although the general treatment outlined earlier is adequate for handling any linear measurement system, some of the instruments warrant separate treatment. Further many salient features would be pointed out while discussing the individual cases.

ZERO ORDER INSTRUMENT

The differential equation that describes the zero order instrument is given by

$$q_0 = \frac{b_0}{a_0} q_i = kq_i,$$

where $k = \frac{b_0}{a_0}$ is the static sensitivity of the instrument. It is obtained by setting all coefficients but a_0 and b_0 in the general differential equation to zero. The block diagram representation of the zero order instrument is given in Fig. 2.7(a).

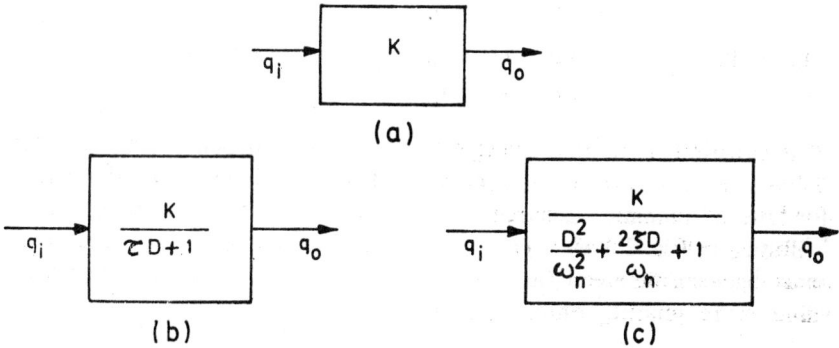

(a)

(b)

(c)

Fig. 2.7 Block diagram representation (a) Zero order (b) First order and (c) Second order instruments

It is obvious from the above equation that no matter how q_i might vary with time. the output follows it perfectly with no time lag and distortion. The dynamic response of a zero order instrument is ideal or perfect. Some examples of this include mechanical level, light-pointer, linear electrical potentiometer, amplifier etc.

FIRST ORDER INSTRUMENT

If all the constants except a_1, a_0 and b_0 in the general differential equation are equal to zero, one obtains

$$\left(\frac{a_1}{a_0} D + 1\right) q_0 = \frac{b_0}{a_0} q_i$$

of $(\tau D + 1)q_0 = kq_i,$

where $\tau \left(= \frac{a_1}{a_0} \right)$ is called time constant and k is the static sensitivity.

An instrument which follows this equation is a first order instrument. The operational transfer function of the first order instrument is, therefore, given by

$$\frac{q_0}{q_i}(D) = \frac{k}{(\tau D + 1)}.$$

The block diagram representation is given in Fig. 2.7(b).

Dynamical response of a first order instrument

(a) *Step response:* In order to study the step response of a first order instrument, the differential equation

$$(\tau D + 1)\, q_0 = k q_i$$

is solved under the following initial conditions:

$$q_i = 0 \text{ for } t < 0$$
$$= q_{is} \text{ for } t \geqslant 0.$$

The solution of this equation, using the usual method. is

$$q_0 = k q_{is}\,(1 - e^{-t/\tau}).$$

This equation shows that the speed of response depends only on the value of τ and is faster if τ is smaller. Thus one desires to minimise τ for faithful dynamic measurements. The normalised output q_0/kq_{is} vs t/τ is plotted in Fig. 2.8(a)   The dynamic error (e_m) at any time is the difference between the ideal (no time lag) value and the actual measured value of the quantity and is given by

$$e_m = q_{is} - q_0/k = q_{is}\, e^{-t/\tau}.$$

The normalised dynamic error e_m/q_{is} vs t/τ plot is shown in Fig. 2.8(b). The error e_m/q_{is} can be obtained from the dynamic response curve by subtracting it from a constant value of unity. It can be seen from Fig. 2.8(a) that the output of a first order instrument, after the application of a step input, rises slowly (depending on its constant) and reaches the final value after a long time. Because of this, dynamic characteristic of the instrument is given by settling time. The settling time is defined as the time required, after the application of the step input, for the output to reach and stay within a specified tolerance band about the final value. As an example 5% settling time means that an instrument requires a time equal to three times its time constant for the output to reach 95% of its final value as shown in Fig. 2 9. If other percentages are used, settling time will have different values. It is, however, obvious that a small settling time is indicative of fast response of the instrument.

(b) *Frequency response:* This requires the solution of the differential equation when an input of the form

$$q_i = A_i \sin \omega t$$

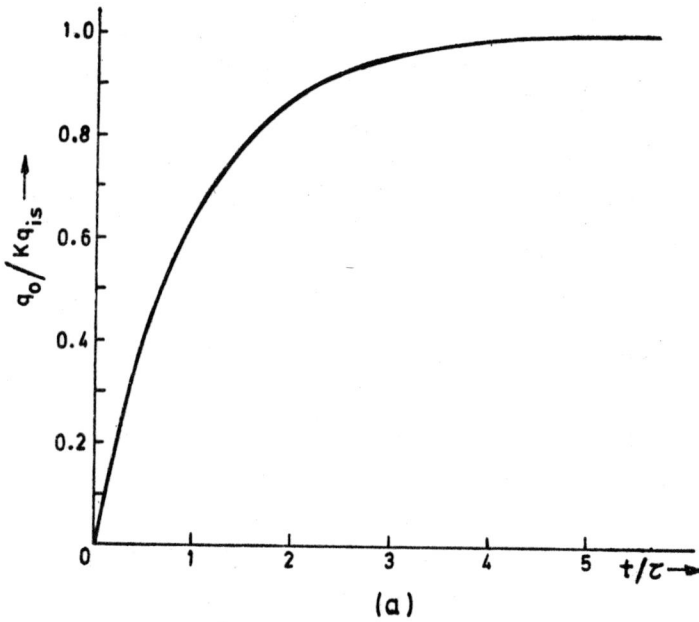

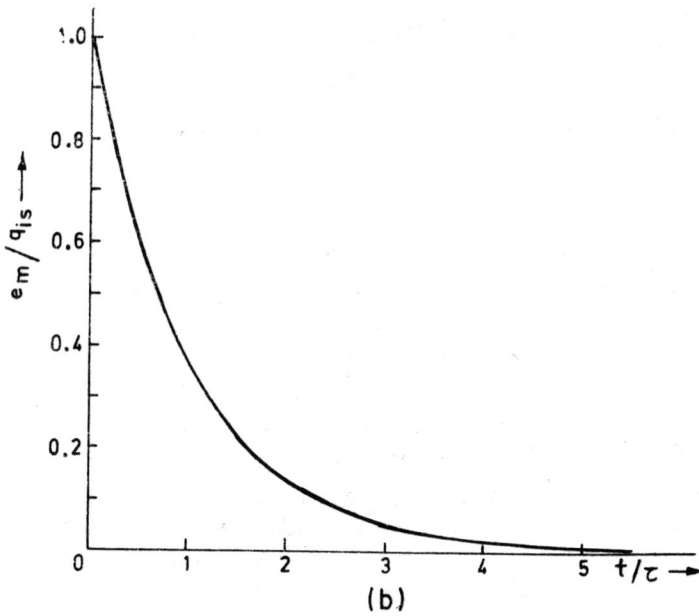

Fig. 2.8 (a) Step function response (b) Error function

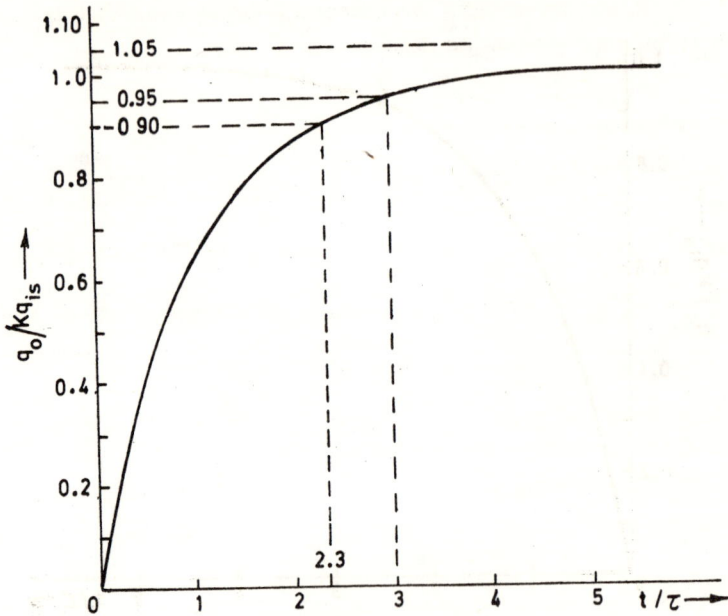

Fig. 2.9   Definition of settling time

is applied.   The initial conditions are:

$$q_i = q_0 = 0 \text{ for } t < 0$$
$$= A_i \sin \omega t \text{ for } t \geqslant 0.$$

The solution, under these initial conditions, is given by

$$q_0 = ce^{-t/\tau} + \frac{kq_i}{1 + i\omega\tau}$$

where c is a constant.   As the time progresses, the complimentary function solution dies out.   The steady state solution therefore is

$$\frac{q_0}{kq_i} = \frac{1}{1 + i\omega\tau} = \frac{1}{\sqrt{1 + \omega^2\tau^2}} \angle -\omega\tau$$

Both the amplitude ratio $\left(\dfrac{1}{1 + \omega^2\tau^2}\right)^{1/2}$ and phase angle $\tan^{-1}(-\omega\tau)$ depend on the product of frequency ω and time constant τ.   The frequency response of a first order instrument would approach an ideal response if the value of $\omega\tau$ is small.   Therefore, there will be an input frequency, for a certain value of τ, below which the measurement would be accurate.   Alternatively the value of τ of the instrument must be sufficiently small for high frequency measurements.

SECOND ORDER INSTRUMENT

If the constants a's and b's except a_2, a_1, a_0 and b_0 are zero, the general differential equation reduces to

$$\left(\frac{a_2}{a_0} D^2 + \frac{a_1}{a_0} D + 1\right) q_0 = \frac{b_0}{a_0} q_i$$

or

$$\left(\frac{D^2}{\omega_n^2} + \frac{2\zeta}{\omega_n} D + 1\right) q_0 = kq_i,$$

where $\omega_n^2 = a_0/a_2 =$ undamped natural frequency, rad/sec,

$$\zeta = \frac{a_1}{2\sqrt{a_0\, a_2}} = \text{damping ratio (dimensionless), and}$$

$k = b_0/a_0 =$ static sensitivity.

An instrument whose performance is governed by this equation is termed as a second order instrument. The operational transfer function of a second order instrument is given by

$$\frac{q_0}{q_i}(D) = \frac{k}{D^2/\omega_n^2 + 2\zeta D/\omega_n + 1}$$

The block diagram representation is given in Fig. 2.7(c). Some examples of second order instruments are force measuring spring scale, spring mass systems under impressed force, L$-$R$-$C circuits, piezoelectric pickups, recorders, etc.

Dynamical response of a second order instrument

(a) *Step response:* To study the step response of a second order system, consider the differential equation

$$\left(\frac{D^2}{\omega_n^2} + \frac{2\zeta D}{\omega_n} + 1\right) q_0 = kq_i$$

with a set of initial conditions

$$q_i = q_0 = 0 \text{ for } t < 0, \text{ and}$$

$$q_i = q_{is} \qquad \text{for } t \geqslant 0$$

Further

$$\frac{dq_0}{dt} = 0 \quad \text{at} \quad t = 0.$$

Three distinct solutions are possible for this equation corresponding to three values of ζ i.e. $\zeta < 1$, $\zeta = 1$ and $\zeta > 1$.

$\zeta < 1$:   UNDERDAMPED SYSTEM

$$\frac{q_0}{kq_{is}} = 1 - \frac{\exp(-\zeta\omega_n t)}{u} \sin(u\omega_n t + \phi),$$

where $u = \sqrt{1 - \zeta^2}$ and $\sin \phi = u.$

$\zeta = 1$: CRITICALLY DAMPED SYSTEM

$$\frac{q_0}{kq_{is}} = 1 - (1 + \omega_n t) \exp(-\omega_n t)$$

$\zeta > 1$: OVERDAMPED SYSTEM

$$\frac{q_0}{kq_{is}} = 1 - \frac{m}{2v} \exp(-g\omega_n t) + \frac{g}{2v} \exp(-m\omega_n t),$$

where $v = \sqrt{(\zeta^2 - 1)}, m = \zeta + v$ and $g = \zeta - v.$

Therefore the response of a second order system to a step input depends on ζ and it is plotted in Fig. 2.10 for different values of ζ. The response is oscillating for smaller values of ζ. It is seen that an increase in the value of ζ reduces the oscillations but results in the slower speed of response as well. A designer can select a proper value of ζ depending on the requirement of settling time. However, the situation is not so simple because the actual form of the input is very complicated, and its actual form influences the best value of ζ as arrived on the basis of settling time. Therefore, for inputs of variable and complicated forms, some compromise should be made. It will be found that most commercial instruments use ζ in the range of 0.6 to 0.7. It will be shown later that this range of ζ values gives good frequency response over the widest frequency range.

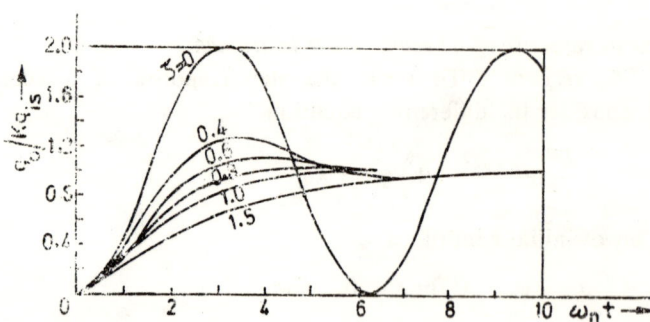

Fig. 2.10 Normalised step response of a second order system

(b) *Frequency response:* When a sinusoidal input of the form $q_i = A_i \sin \omega t$ is applied to the system, a steady state solution of the differential equation represents a sinusoidal output with a different amplitude and phase shift but having the same frequency. The sinusoidal transfer function of a second order instrument is expressed as

$$\frac{q_0}{q_i}(i\omega) = \frac{k}{\left(\dfrac{i\omega}{\omega_n}\right)^2 + 2\zeta\left(\dfrac{i\omega}{\omega_n}\right) + 1}$$

This can be put in the form

$$\frac{q_0/k}{q_i}(i\omega)=\frac{1}{\left[\left\{1-\left(\dfrac{\omega}{\omega_n}\right)^2\right\}^2+4\zeta^2\left(\dfrac{\omega}{\omega_n}\right)^2\right]^{1/2}}\angle\phi$$

where $\qquad \tan\phi=\dfrac{2\zeta}{\left(\dfrac{\omega}{\omega_n}-\dfrac{\omega_n}{\omega}\right)}.$

Figure 2.11(a) illustrates the plots of q_0/kq_i vs ω/ω_n for various values of ζ. The resonance occurs at $\omega=\omega_n$ i.e. where the forcing frequency ω is equal to the natural frequency ω_n of the system. Very large amplitudes will be reached unless either damping is introduced or the system operates at $\omega=\omega_n$ for a very short duration. Further as ζ increases, the range of frequencies, for which amplitude ratio curve is relatively flat, also increases. The amplitude ratio curves remain fairly flat for the widest range of frequencies only for ζ of about 0.6 to 0.7. While zero phase angle is ideal, it is rarely possible to realise it even approximately. The variation of phase angle ϕ with ω/ω_n is shown in Fig. 2.11(b).

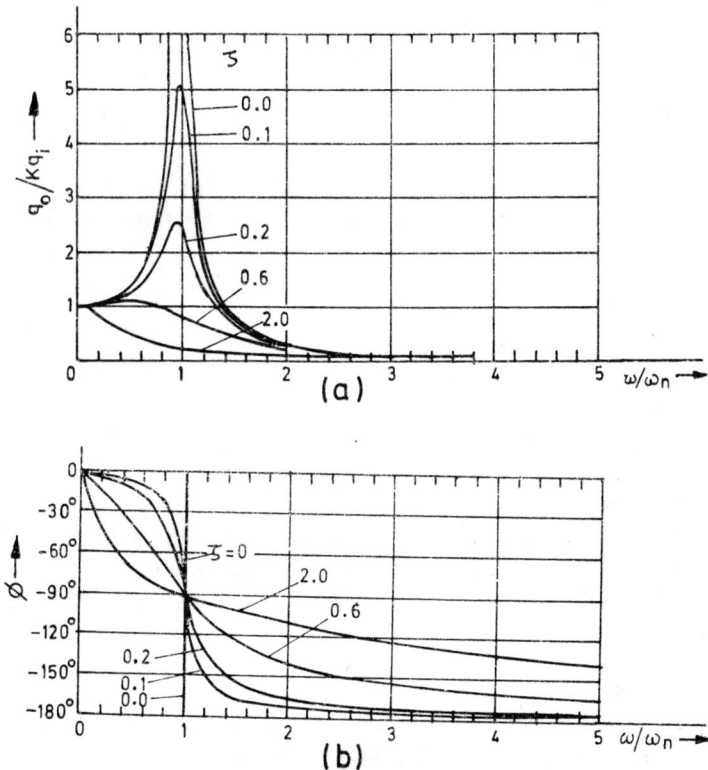

Fig. 2.11   Frequency response of second order system

2.6   Experimental determination of system parameters

The methods to determine the order of an instrument, and to measure the parameters which characterise it are now discussed with respect to the two elementary input i.e. step and sinusoidal inputs.

STEP INPUT METHODS

Zero order instrument

It is seen that a zero order instrument is a perfect instrument and its dynamical response is ideal. It has no time lag or distortion. The constant k, the static sensitivity, is the only parameter that characterises it. It can be determined by the process of static calibration.

First order instrument

A first order instrument is characterised by two parameters, τ and k. The static sensitivity k is obtained by static calibration. For the determination of time constant τ, the following two methods, which employ step input, are described:

(i)   One common method is to apply a step input and measure τ as the time required to reach 63.2% of the final value.

Analytically

$$\frac{q_0/k}{q_{is}} = 1 - e^{-t/\tau} = 0.632$$

or

$$e^{-t/\tau} = 0.368 = e^{-1}.$$

Thus

$$t = \tau.$$

The method requires measurement only at two points, i.e. at $t = 0$ and $t = \tau$ and is influenced by the inaccuracies in the determination of $t = 0$ point. Further, it does not determine the order of the instrument.

(ii)   In the second method also, a step input is applied. Therefore, the response of the first order instrument is described by

$$1 - \frac{q_0/k}{q_{is}} = e^{-t/\tau}$$

Let $z = \log_e(1 - q_0/k\, q_{is})$ where z is the $\log_e$ of the incomplete response of the system. A plot of z vs t must be a straight line if the system is of first order. The slope of this line is $-1/\tau$. This gives a more accurate value of τ because a best straight line is drawn through all the points. The method is heavily dependent on the initial few measurements, particularly below τ value. Both these methods are graphically described in Fig. 2.12 (a) and (b).

Second order instrument

The second order instrument is characterised by three parameters, k, ζ and ω_n. The value of k is again obtained by static calibration, while the methods to determine values of ζ and ω_n depend on the value of ζ itself.

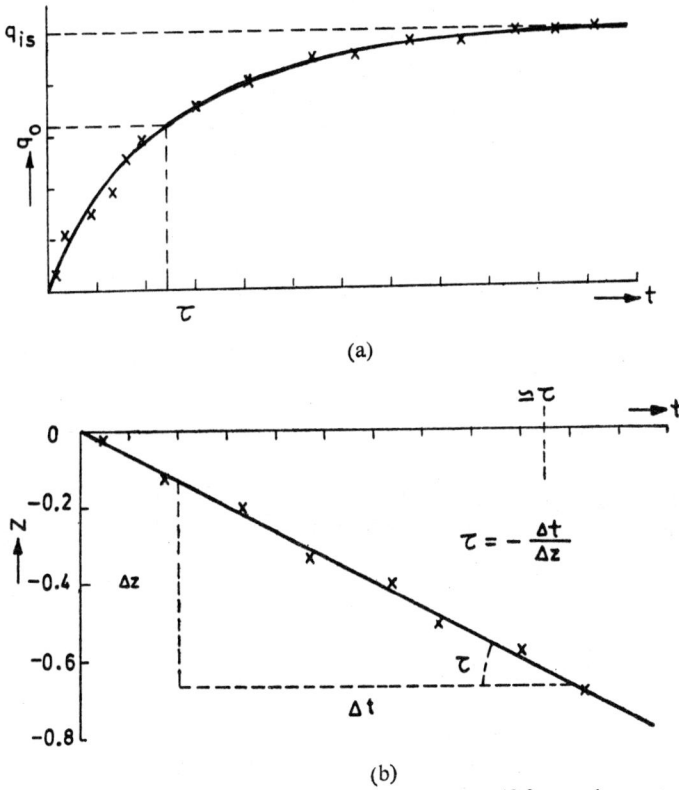

(a)

(b)

Fig. 2.12   Measurement of τ from step response of first order system

(i) $\zeta < 1$: UNDERDAMPED

The step response of a second order under damped ($\zeta < 1$) instrument is given by

$$q_0 = A\left[1 - \frac{\exp\left(-\zeta\omega_n t\right)}{u} \sin\left(u\omega_n t + \phi\right)\right]$$

where $A = k q_{is}$.

The output q_0 as a function of time is shown in Fig. 2.13.   If the time period of damped oscillations is T, then

$$u\omega_n t + \phi + 2\pi = u\omega_n (t + T) + \phi$$

which gives $\omega_n = \dfrac{2\pi}{uT} = \dfrac{2\pi}{T\sqrt{1 - \zeta^2}}$.

Therefore to obtain the value of ω_n, the value of u and hence of ζ is required.   If the overshoot is a, then

$$A + a = A\left[1 + \frac{\exp\left(-\zeta\omega_n t\right)}{u}\right].$$

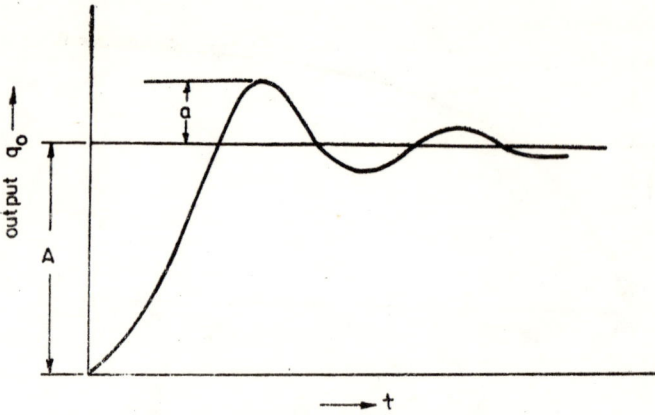

Fig. 2.13 Step response of a second order system

Exponential term will vary faster than the denominator u with ζ, and under the assumption that ζ is fairly small, the denominator u can be approximated to unity.

Therefore

$$\frac{a}{A} = \exp\left(-\zeta\omega_n t\right).$$

This equation holds good when

$$\sin\left(u\omega_n t + \phi\right) = -1 \text{ with } \sin\phi = u \simeq 1,$$

or

$$\omega_n t = \frac{\pi}{u}.$$

Therefore

$$\frac{a}{A} = \exp\left(-\zeta\frac{\pi}{u}\right) = \exp\left(-\frac{\zeta\pi}{\sqrt{1-\zeta^2}}\right).$$

This gives

$$\zeta = \frac{1}{\left[1 + \left(\pi/\ln\dfrac{a}{A}\right)^2\right]^{1/2}}$$

This value of ζ is used to obtain the value of ω_n from the relation

$$\omega_n = \frac{2\pi}{T\sqrt{1-\zeta^2}}.$$

However, when the system is lightly damped there is really no need for a step input. Any fast transient input will produce an oscillatory response similar to the one shown in Fig. 2.14. The value of ζ can be approximated by the formula

$$\zeta = \frac{\ln\left(x_1/x_n\right)}{2\pi n},$$

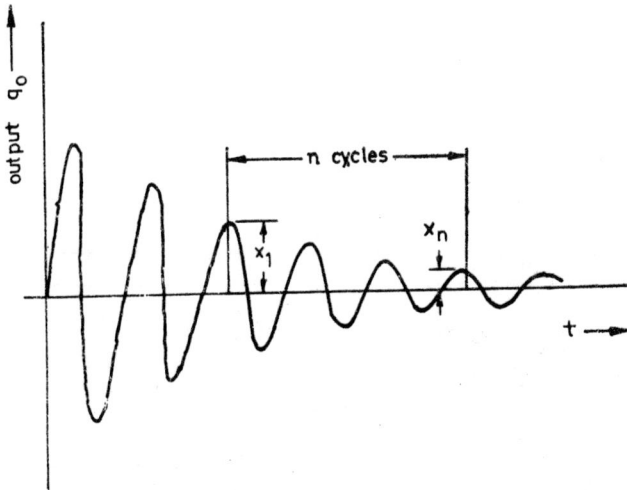

Fig. 2.14   Response of a second order system to a transient input

where x_1, x_n and n are defined in Fig. 2.14.   This formula is quite accurate when $\zeta < 0.1$.   This value of ζ is used to obtain the value of ω_n. With in the validity of the approximation, the value of ζ does not depend on n provided the system is second order linear with $\zeta < 0.1$.

(ii) $\zeta > 1$:   OVERDAMPED SYSTEM

The response of an overdamped system to a step input does not show any oscillations and hence the determination of ζ and ω_n become very difficult.   To overcome this disadvantage the response of the system is expressed by two time constants τ_1 and τ_2 instead of ζ and ω_n. The time constants τ_1 and τ_2 are defined as

$$\tau_1 = \frac{1}{(\zeta - \sqrt{(\zeta^2 - 1)})\,\omega_n} \quad \text{and} \quad \tau_2 = \frac{1}{(\zeta + \sqrt{(\zeta^2 - 1)})\,\omega_n}.$$

Evidently $\tau_1 > \tau_2$ as $\zeta > 1$.   The response of a second order instrument to a step input can be expressed in terms of τ_1 and τ_2 as

$$\frac{q_0/k}{q_{is}} = 1 + \frac{\tau_1}{\tau_2 - \tau_1}\,e^{-t/\tau_1} - \frac{\tau_2}{\tau_2 - \tau_1}\,e^{-t/\tau_2}.$$

To find τ_1 and τ_2 and hence ζ and ω_n, from the step response curve, following procedure is adopted:

(1)   From the step response of a second order instrument shown in Fig. 2.15(a), 'percentage incomplete response' R_p is obtained according to

$$R_p = \left[1 - \frac{q_0/k}{q_{is}}\right] \times 100$$

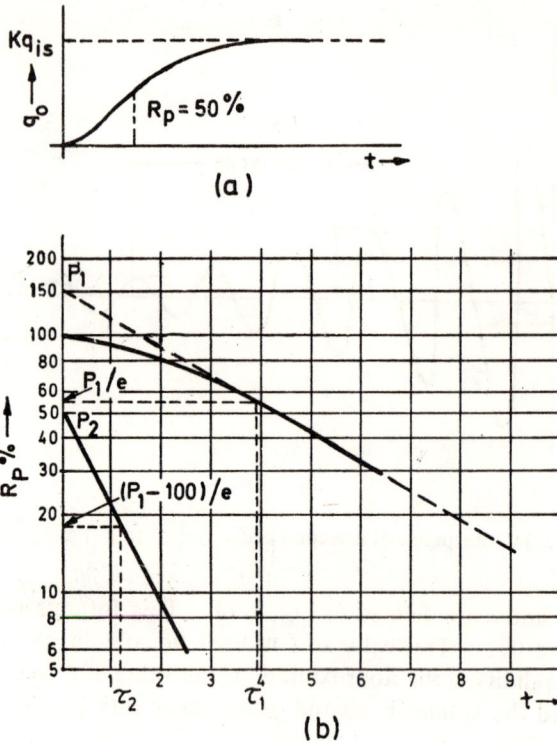

Fig. 2.15   Step test for over damped second order system

(2)   A curve is plotted between R_p on a logarithmic scale and t on a linear scale.   Fig. 2.15(b) shows a plot of R_p vs t.   For a second order system and for large values of t, this curve tends to a straight line.   The intercept of this straight line (extended to $t = 0$) on the R_p axis is denoted by P_1.   The time constant τ_1 is the time measured on the time axis on this straight line for a value of $R_p = P_1/e = 0.368\ P_1$.

(3)   Fig. 2.15 (b) shows another curve which is the difference between the straight line asymptote and R_p.   This curve must be a straight line if the system is second order.   The time corresponding to a value $(P_1 - 100)/e$ for R_p on this straight line is numerically equal to τ_2.

The validity of the results thus obtained can be checked by the following mathematical procedure :

(1)   By definition

$$R_p = \left[\frac{\tau_2}{\tau_2 - \tau_1} e^{-t/\tau_2} - \frac{\tau_1}{\tau_2 - \tau_1} e^{-t/\tau_1} \right] \times 100.$$

(2)   Thus $\ln R_p = \ln 100 - \dfrac{t}{\tau_1} + \ln \left[\dfrac{\tau_2}{\tau_2 - \tau_1} e^{-t(\tau_1 - \tau_2)/\tau_1 \tau_2} - \dfrac{\tau_1}{\tau_2 - \tau_1} \right]$

Since $\tau_1 > \tau_2$ the term $\exp\left[-t(\tau_1 - \tau_2)/\tau_1\tau_2\right]$ will decrease fast with time and would vanish for large t. So for large t,

$$\ln R_{ps} = \ln 100 - \frac{t}{\tau_1} + \ln \frac{\tau_1}{\tau_1 - \tau_2}.$$

The relation between $\ln R_{ps}$ and t is linear. R_{ps} is the value of R_p corresponding to straight line asymptote. R_{ps} can be rewritten as

$$R_{ps} = 100 \; \frac{\tau_1}{\tau_1 - \tau_2} \; e^{-t/\tau_1} = P_1 e^{-t/\tau},$$

where P_1 is the value of R_{ps} at $t = 0$. Therefore at $t = \tau_1$,

$$R_{ps} = P_1/e.$$

(3)   The difference between R_p and R_{ps} can be obtained from the equations already given and is

$$\Delta R = R_{ps} - R_p$$

$$= 100 \frac{\tau_2}{\tau_1 - \tau_2} \; e^{-t/\tau_2}$$

The relation between $\ln \Delta R$ and t is linear for all values of t. The value of ΔR at $t = 0$ is equal to

$$\Delta R \Big|_{t=0} = 100 \frac{\tau_2}{\tau_1 - \tau_2} = P_2 = P_1 - 100.$$

Thus

$$\Delta R = (P_1 - 100) \; e^{-t/\tau_2}.$$

At $t = \tau_2$, the value of ΔR should be P_2/e or $(P_1 - 100)/e$ From these values of τ_1 and τ_2 the values of ζ and ω_n can now be calculated.

The dynamic behaviour of an instrument does not depend on the magnitude or sign of the step input. The step response curves obtained experimentally for different step inputs must be same. If the tests conducted using step inputs of different magnitudes and sign produce results which are different both quantitatively and qualitatively, it is an indication of non-linear response of the instrument.

FREQUENCY RESPONSE METHODS

The frequency response methods are very simple and easy to perform but are very expensive due to the non-availability of mechanical sine generators. Often impulse input response is used to obtain frequency response.

First order instrument

The frequency response of a first order instruement is given by

$$\frac{q_0}{q_i}(i\omega) = \frac{k}{\sqrt{1 + \omega^2\tau^2}} \; \angle\phi \text{ where } \tan\phi = -\omega\tau.$$

The amplitude ratio $k/\sqrt{1 + \omega^2 \tau^2}$ and phase ϕ are plotted on logarithmic scale as a function of $\log \omega$ in Fig. 2.16.

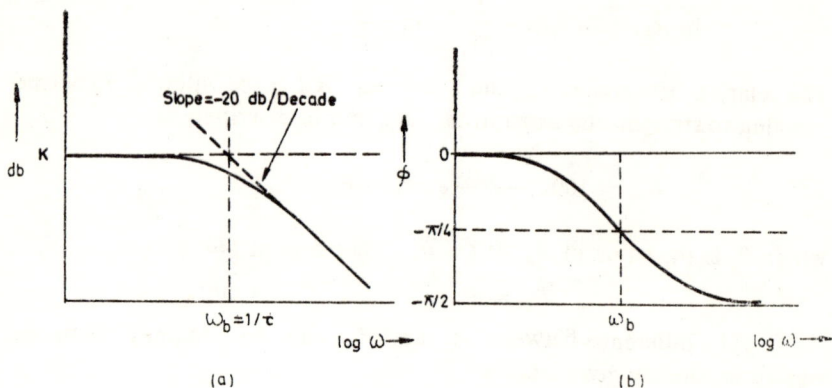

Fig. 2.16　Frequency response test of a first order system (a) Amplitude in db vs log ω, and (b) Phase vs log ω

At the low frequency side $\omega\tau < 1$ and hence $\dfrac{q_0}{q_i} = k$. While on the high frequency side $\omega\tau > 1$ and $\dfrac{q_0}{q_i} = \dfrac{k}{\omega\tau}$. Therefore the asypmtote to the curve at high frequency side makes an angle of $-$ 20db/decade. It thus suggests that the instrument is first order if the asymptote makes an angle of $-$ 20db/decade. It is also seen that the asymptotes to the low and high frequencies side of the curve intersect at a point which corresponds to a frequency ω_b such that $\omega_b = 1/\tau$. Thus the value of τ can be calculated from the experimentally obtained value of ω_b. Further more the phase angle at ω_b is $-45°$.

Second order instrument

The frequency response of a second order instrument is given by

$$\frac{q_0}{q_i}(i\omega) = \frac{k}{\left[\left(1 - \left(\dfrac{\omega}{\omega_n}\right)^2\right)^2 + 4\zeta^2\left(\dfrac{\omega}{\omega_n}\right)^2\right]^{1/2}} \angle \phi,$$

where　　$\tan\phi = \dfrac{2\zeta}{\left(\dfrac{\omega}{\omega_n} - \dfrac{\omega_n}{\omega}\right)}$. The frequency response depends on both

ζ and ω_n. The cases of $\zeta < 1$, $\zeta = 1$ and $\zeta > 1$ are discussed here.

(i)　$\zeta < 1$: UNDERDAMPED

The variation of amplitude ratio (output) with ω/ω_n under this case is shown in Fig. 2.17. The curve shows a maximum at $\omega = \omega_p = \omega_n (1 - 2\zeta^2)^{1/2}$. The amplitude ratio at ω_p is $k/2\zeta (1 - \zeta^2)^{1/2}$. Therefore the ratio of the amplitude ratios at $\omega = 0$ and $\omega = \omega_p$ can be written as

$$\frac{(q_0/q_i)\big|_{\omega=0}}{(q_0/q_i)\big|_{\omega=\omega_p}} = 2\zeta(1-\zeta^2)^{1/2}.$$

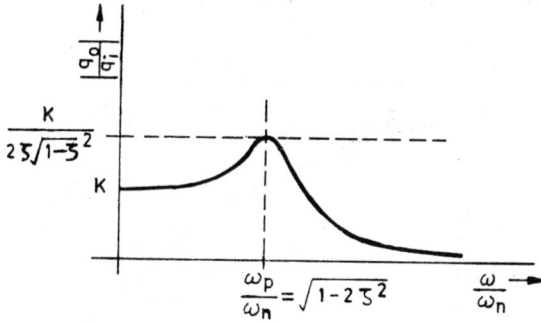

Fig. 2.17 Frequency response test for a second order system
(underdamped)

The value of ζ is thus obtained from the experimentally measured values of amplitude ratios at $\omega = 0$ and $\omega = \omega_p$ and the value of ω_n from the measurement of frequency at which the hump in the curve occurs.

Instead of plotting output vs frequency, one can also plot db vs log ω. Figure 2.18 shows such a plot of db vs log ω. The slope of the straight line drawn asymptotically to the curve at high frequency side is -40 db/decade. The intersection of asymtotic lines at low and high frequency sides of the curve gives log ω_n as shown in Fig. 2.18. This method is applicable for both the underdamped and critically damped systems. This however gives the value of ω_n for underdamped system.

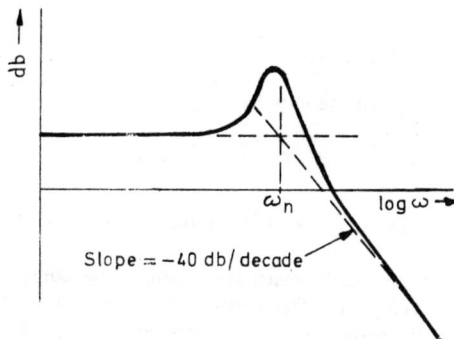

Fig 2.18 Frequency response test for second order system
(underdamped and critically damped)

(ii) $\zeta > 1$: OVERDAMPED

The plot of db vs log ω for an overdamped system is shown in Fig. 2.19. The curve distinctly has three asymptotes, at low, medium and high frequency sides, and hence two intersection points. The intersection points of the asymptotes occur at the break frequencies ω_1 and ω_2, where $\omega_1 = 1/\tau_1$, and $\omega_2 = 1/\tau_2$. The time constants τ_1 and τ_2 are related to ζ and ω_n for over damped system as discussed earlier and hence can be computed from the values of τ_1 and τ_2 obtained from measurements.

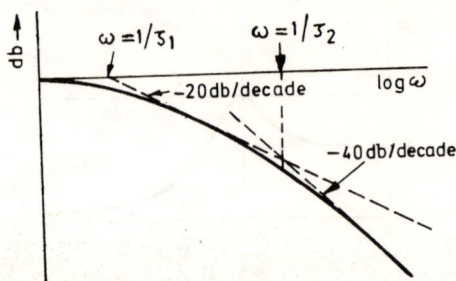

Fig. 2.19 Frequency response test for a second order system
(overdamped)

Exercises

1. A ramp input is applied to a first order instrument. Show that its response is given by

$$\frac{q_0}{k\dot{q}_{i,s}\tau} = \frac{t}{\tau} - (1 - e^{-t/\tau}).$$

Also show that the instrument reads what the input was τ seconds before.

2. Prove that the impulse response of a first order instrument is given by

$$q_0 = k\frac{A}{\tau} e^{-} \quad ,$$

where A is the strength of the impulse defined as

$$A = \int_0^T u(t)\,dt : T \to 0,\ u(t) \to \infty.$$

Also show that only the strength of the impulse and not its shape effects the response.

3. A thermocouple is a first order instrument with a time constant τ. A part of a continuous sinusoidal input of frequency ω is applied to it. Assuming ω to be very small, obtain the responce of the thermocouple. If a single temperature pulse of sinusoidal shape (halfwave) of duration t is applied to this thermocouple, what will be its response. The duration of the pulse is four times the time constant. Compare these two responses.

4. An input temperature pulse of maximum temperature T_0 and shape as shown in Fig. 2.20 is applied to a thermocouple of time constant τ. The duration of the pulse is 7τ. Obtain the output of thermocouple as a function of time. If the duration of this pulse is $\tau/10$ instead of 7τ, what difference in output is observed?

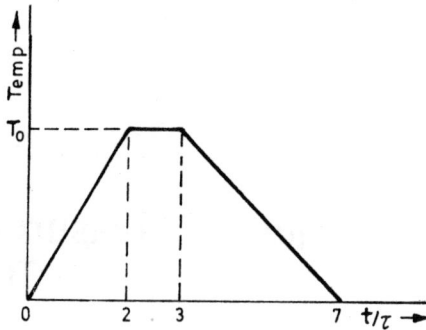

Fig. 2.20   Temperature input to a thermocouple

5. A parabolic input of the form $q_i = q_{is}t^2$ is applied to a first order instrument. Show that its response is given by

$$\frac{q_0}{kq_{is}\tau^2} = \frac{t^2}{\tau^2} - 2\,(e^{-t/\tau} + \frac{t}{\tau} - 1).$$

6. Show that the ramp response of a second order instrument is given as follows:
   (i) underdamped: $\zeta < 1$

$$\frac{q_0}{kq_{is}} = t - \frac{2\zeta}{\omega_n}\,[1 - \frac{\exp\,(-\zeta\omega_n t)}{2\zeta\sqrt{1-\zeta^2}}\,\sin\,(\sqrt{(1-\zeta^2)}\,\omega_n t + \phi)],$$

where

$$\tan \phi = \frac{2\zeta\sqrt{1-\zeta^2}}{2\zeta^2 - 1}.$$

   (ii) critically damped: $\zeta = 1$

$$\frac{q_0}{kq_{is}} = t - \frac{2}{\omega_n}\left[1 - \left(1 + \frac{\omega_n t}{2}\right)\exp\,(-\omega_n t)\right]$$

   (iii) overdamped: $\zeta > 1$

$$\frac{q_0}{kq_{is}} = t - \frac{\zeta}{\omega_n}\left[2 + \left(\frac{2\zeta^2 - 1}{2\zeta\sqrt{\zeta^2 - 1}} - 1\right)\exp - (\zeta - \sqrt{\zeta^2 - 1})\,\omega_n t\right.$$

$$\left. - \left(\frac{2\zeta^2 - 1}{2\zeta\sqrt{\zeta^2 - 1}} + 1\right)\exp - (\zeta + \sqrt{\zeta^2 - 1})\,\omega_n t\right]$$

Inaccuracy of Measurements and Its Analysis

3.1 Introduction

The purpose of any measurement is to describe some physical property of an object or a system quantitatively, viz., its length, temperature, pressure, etc. Every measurement of such a quantity has a certain amount of uncertainty. This may be explained by considering following three different experiments for the measurement of the diameter of a disc:

(1) First, three different instruments are chosen: a caliper square, a micrometer screw and an Abbe metroscope. Each of these instruments may give different measurements for the diameter of the same disc.

(2) Next, three instruments of the same type are used, for example three metroscopes to measure the diameter of the same disc. Again three values of the diameter by the three different instruments may be obtained.

(3) Lastly, only one instrument is used to measure the diameter of the same disc three times. Here as well one may get three different values.

Thus one finds that even repeated measurements by the same instrument may give different values of the same physical quantity. Therefore, it can be concluded that all measured values are inaccurate to some degree. It is in fact impossible to find the true value of a physical quantity, although it is safe to assume that the true value exists. The aim is to find the most probable value and assign an uncertainty to it. In other words, the aim of the experimenter need not be to make the uncertainty of his measurements as small as possible, a cruder result may serve his purpose but he must be assured that the uncertainty in his measurements is so small as not to affect the conclusions he draws from his results. Therefore the determination of the uncertainty of a measurement and a consistant way of specifying the uncertainty in an analytical fashion is a necessary part of the act of measurement.

3.2 Types of errors

The errors, in general, are classified as *accidental* and *systematic*. Usually each error gives an accidental and a systematic component which should be used in assigning uncertainty to the measured value. Besides there may be a *gross blunder* that arises due to faulty design, faulty circuit etc. The experimenter should be able to locate and eliminate this hopefully.

ACCIDENTAL ERROR

The accidental errors are random in their occurrence and variable in magnitude, and usually follow a certain statistical law—a normal distribution law. A measure of accidental errors is, therefore, standard deviation of the distribution. Standard deviation on either side of the maximum of a normal distribution encloses 68.27% of the area under the curve. With increasing number of measurements of the same quantity, the value of standard deviation gets smaller and the value of the quantity approaches its true value.

SYSTEMATIC ERROR

The systematic errors have definite magnitude and direction. These are usually more troublesome as repeated measurements need not necessarily reveal them. Even when their existence or nature has been established, it is sometimes very difficult to determine and eliminate them. Therefore in most of the cases the systematic errors are not used for correcting the measurements but are taken with their full values along with accidental errors as uncertainty.

The experimenter may sometimes use theoretical methods to estimate the magnitude of systematic errors. For instance the error introduced in the measurement of temperature due to exposed portion of a mercury thermometer or the error introduced in the measurement of length due to temperature change can be estimated theoretically.

3.3 Sources of error

1. *Instrument calibration*

Due to frequent use of a measuring instrument and also of aging, the instrument may go out of calibration. The measurements made with such an instrument will be in error; this type of error usually is regular and may be called systematic. A comparison with a standard instrument enables such uncertainty to be accounted for. An instrument should therefore be sent for calibration at frequent intervals.

2. *Instrument reproducibility*

Even if an instrument has been calibrated under a set of conditions, the measurements may still be in error if the instrument is not being used under conditions identical to those prevailing during calibration

Mechanical defects like sleekness and friction in metal bearings, backlash in micrometer screws etc. lead to errors in measurement. These errors could be either accidental or systematic.

3. *Measuring arrangement*

The measuring arrangement employed may sometimes influence the measurements. This is particularly so when the comparator law of Abbe is not strictly followed while measuring length. According to this law, errors of first order are avoided when the measuring instrument and scale axes are collinear. Strict adherence to this law for designing and usage of an opto-mechanical instrument is essential. Similar example can be found in other areas.

4. *Workpiece*

The nature of workpiece viz hardness, roughness etc. may lead to errors in measurement. Most mechanical and opto-mechanical instruments contact the workpiece under constant pressure condition. Therefore hard and soft workpieces would respond differently and this may lead to errors in measurement.

5. *Environmental conditions*

Environmental conditions such as humidity, pressure, temperature, electrical or magnetic field also influence the measurements if the instrument is not used under conditions prevailing during calibration. A very well known example is in the area of length measurement where a workpiece of nominal length L is measured with an instrument calibrated at 20°C. The error introduced when the measurement is done at temperature t will be

$$= (a_2 - a_1)L(t - 20),$$

where a_1 and a_2 are the thermal expansion coefficients of workpiece and of scale of the measuring instrument respectively.

6. *Observer's skill*

Human element, despite of automation, often enters into a measurement scheme and plays a decisive role. Therefore observer's skill is to be considered in the act of measurement. It is an established fact that the measurement data of a physical quantity varies from one observer to another, and even for the same observer it may vary with his physical and mental states. Such errors may be of both systematic and random nature depending on the conditions of the experiment.

3.4 Mean as a best value

When a large number of measurements on a physical quantity are made, all the measurements may differ in magnitude from each other and also from the true value that is not known. The task is to obtain the value of the physical quantity which is very close to its true value. It would be shown that the mean is the best value.

If a quantity with the assumed true value x_0 (unknown) is measured n times and the observations are recorded as $x_1, x_2, \ldots x_n$ units, then $x_r = x_0 + e_r$, where e_r is the uncertainty in the r_{th} observation and can take both positive and negative values. The arithmatic mean $\bar{x}$ of the n measurements is

$$\bar{x} = \frac{\Sigma x_r}{n} = x_0 + \frac{\Sigma e_r}{n}$$

Since some of the errors are positive and some are negative, the term $(\Sigma e_r/n)$ will be very small. In any case it will be numerically smaller than the largest value of the separate errors. Thus if e is the largest numerical value of an error in any of the n measurements then,

$$\frac{\Sigma e_r}{n} \ll e$$

and consequently

$$\bar{x} - x_0 \ll e.$$

Hence, in general, $\bar{x}$ will be nearer to x_0 and may be taken as the best value of a physical quantity. As a rule, larger the value of n, nearer $\bar{x}$ approaches x_0.

It may be kept in mind that it is not possible to find the values e_r as x_0 is unknown. It is, thus, usual to examine the scatter or dispersion about the mean value $\bar{x}$ rather than x_0. Therefore x_r can be written as

$$x_r = x_0 + e_r = \bar{x} + d_r$$

It follows that $\Sigma \, d_r = 0$.

3.5 Accuracy and precision

Accuracy refers to the agreement of the result of measurement with the true value of the measured quantity. If an instrument measures x instead of X then error, $(X - x)$, or percentage error $(X - x) \, 100/X$ is a measure of accuracy of the instrument. The nearness of the measurement to its true value is the accuracy. However the true value is not known. Hence the most probable value or the mean is used and the accuracy is defined, generally as a percentage.

An experimenter should know the degree of accuracy that can be achieved from an instrument. There are two methods adopted for specifying the accuracy of an instrument. In one method, accuracy is expressed as a percentage of the full scale reading of the instrument. As an example, a pressure gauge having a range of 0 to 1 Kgf/cm² is quoted an accuracy of $\pm 1\%$ of the full scale, i.e. no error greater than 0.01 kg/cm² can be expected for any reading that might be taken with this gauge. It may be noted that for an actual reading of 0.1 Kgf/cm², an error of 0.01 Kgf/cm² is 10% of the reading. Another method gives the error as a percentage of any reading with a qualifying statement to apply to the lower end of the scale. For example, an accuracy of $\pm 0.5\%$ of the reading or ± 0.05 Kgf whichever is greater may be quoted for a spring scale.

Precision refers to the repeatability of a measuring process, i.e. the closeness with which the measurement of the same physical quantity agree with one another regardless of any systematic error. Mathematically if d_r is small, the precision is high and e_r is small, the accuracy of measurement is high. The difference between the terms 'accuracy' and 'precision' can be illustrated by considering an example: a known voltage of 100 volts is measured 5 times by a certain instrument. The indicated values are 104, 103, 105, 103, 105 volts. It is obvious that the instrument cannot be depended upon for an accuracy better than 5%, while it has a precision of 1%. It should be noted that the accuracy of an instrument can be improved by calibration but not beyond the precision. Accuracy, therefore, includes precision but the converse is not necessarily true.

3.6 Statistical analysis of data

DISTRIBUTION CURVES

A graphical presentation of data has a visual appeal and provides quick assimilation of information contained in the raw data. One of the ways to graphically present the data is by drawing a 'histogram'. A method to draw a histogram is illustrated with the help of an example. Table 3.1 gives the percentage of marks obtained in a particular paper

TABLE 3.1 No. of students in the class = 100

S. No.	Interval of marks	frequency
1	0– 9	1
2	10–19	3
3	20–29	4
4	30–39	11
5	40–49	20
6	50–59	29
7	60–69	22
8	70–79	8
9	80–89	1
10	90–99	1
	Total	100

by the students. The marks are grouped in a convenient interval of a variable (marks) and the number of data in each interval is called the frequency.

Suppose a quantity z is defined by

$$z = \frac{\dfrac{\text{Number of measurement}}{\text{data in an interval}} \bigg/ \dfrac{\text{Total number of}}{\text{measurements}}}{\text{Width of interval}}$$

and a graph with height z for each interval is plotted. Such a graph, that contains a series of rectangles is called a 'histogram'. The area under every rectangle is proportional to frequency in that interval. If the widths of the intervals are equal, the heights of the rectangles are proportional to the frequencies. A histogram for the data of Table 3.1 is plotted in Fig. 3.1. If the center points of all the rectangles are sequentially joined, another graphical representation, called 'frequency polygon' results (Fig. 3.1).

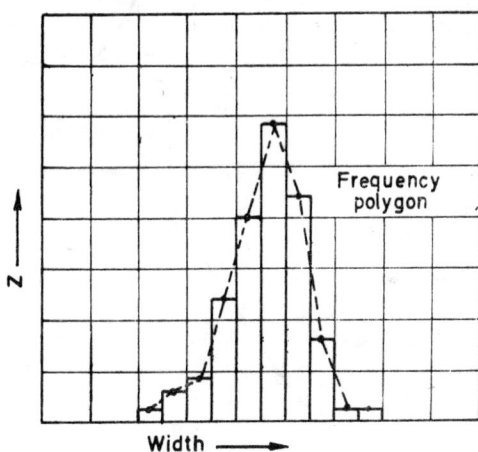

Fig. 3.1 Histogram and frequency polygon

The area, the product of z and width of the interval, under each interval is numerically equal to the probability that a particular measurement will fall in that interval. When an infinite number of measurements are made and the intervals are made as small as possible, the polygon would approach a smooth curve. If this limiting case is taken as a mathematical model of a real physical situation, the function $z = f(x)$ is called the probability density function for the mathematical model of a real physical process. From the definition of $f(x)$,

$$\int_{-\infty}^{\infty} f(x)dx = 1.$$

The probability of a measurement x lying between a and b [Fig. 3.2(a)] will be given by

$$p(a<x<b) = \int_a^b f(x)dx$$

The probability information is sometimes given in terms of the cumulative distribution function $F(x)$, which is defined as the probability that the measurement is less than any chosen value of x i.e. $F(x) = \int_{-\infty}^x f(x)\ dx$. This is illustrated in Fig. 3.2(b).

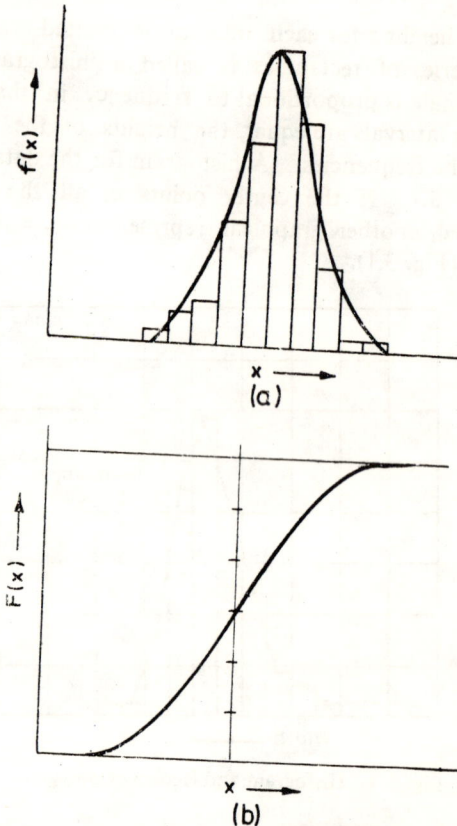

Fig. 3.2 (a) Probability distribution function and
(b) Cumulative distribution function

THE MEAN, THE MEDIAN AND DISPERSION

The mean

Assume that a physical quantity is measured n times. The recorded measurements of this are $x_1, x_2, \ldots x_r \ldots$ which occur with frequencies $f_1, f_2, \ldots f_r, \ldots$ respectively such that $f_1+f_2+\ldots+f_r+\ldots = \Sigma f_i = n$. The mean value of the measurement is defined as

$$\bar{x} = \frac{\Sigma f_i\ x_i}{n}$$

Evaluation of $\bar{x}$ using this relation may sometimes be laborious. This may be simplified by introducing the concept of an assumed mean m_0, a constant.

The variable x_i is expressed in terms of a new variable x_i' as

$$x_i = x_i' + m_0.$$

Thus

$$\Sigma f_i x_i = \Sigma f_i (x_i' + m_0) = \Sigma f_i x_i' + nm_0.$$

Hence

$$\bar{x} = \bar{x}' + m_0,$$

where $\bar{x}'$ is the mean of the new variable x_i'.

Often scaling is also done simultaneously. If the scale factor is F_s, a new variable x_i' can be defined such that $x_i = F_s x_i' + m_0$. This gives $\bar{x} = F_s \bar{x}' + m_0$. The simplicity that is provided by scaling and assumed mean will become evident, when an example is considered.

The median

If a set of measurements are arranged in ascending or descending order of magnitude, the measurement in the middle of the set is called the median.

Dispersion

A very important characteristic of a set of data is its dispersion or scatter about some value, say mean value. Various parameters are employed to measure the dispersion i.e., the range, the mean deviation and the standard deviation.

The range: The range of a frequency distribution is defined as the difference between the least and greatest values of the variable. The range includes all data, even those data which may occur very infrequently.

The mean deviation: It has been shown that the sum of deviations from the mean value is zero, i.e. $\Sigma d_r = 0$. The mean deviation of a set of data is defined as the mean of the numerical values of the deviations from the mean, i.e.

$$\bar{d} = \frac{\Sigma f_i \mid d_i \mid}{n}$$

The standard deviation: The variance, s^2, of a set of data is defined by

$$s^2 = \frac{\Sigma f_i (x_i - \bar{x})^2}{\Sigma f_i}$$

The square root of the variance is called the standard deviation s. The standard deviation is the root mean square deviation of the data measured from the mean. However, when the number of observations is not very large, a standard deviation s' is defined as

$$s' = \sqrt{\frac{\Sigma f_i (x_i - \bar{x})^2}{n - 1}}$$

The number n is replaced by $n-1$ in the denominator.   This is known as Bessel's correction.

When the calculations are made with a scaled up and shifted variable x'_i, defined earlier, the standard deviations is given by

$$s = F_s \sqrt{\frac{\Sigma f_i \, (x'_i - \bar{x}')^2}{n}}$$

GAUSSIAN DISTRIBUTION

As has been said earlier that the measured values are always uncertain and this uncertainty may be due to both the systematic and random errors. The random errors follow a statistical distribution and generally it is a normal or Gaussian distribution. It is represented by

$$f(x) = \frac{1}{\sqrt{2\pi\sigma^2}} \; \exp\left[-(x - \mu)^2/2\sigma^2\right]$$

where μ and σ are the mean and standard deviation respectively of the distribution and are defined by

$$\mu = \frac{\Sigma f_i x_i}{n} \; ; \quad n \to \infty;$$

and

$$\sigma^2 = \frac{\Sigma f_i \, (x_i - \mu)^2}{n} \; ; \quad n \to \infty .$$

An estimate of μ and σ is given by $\bar{x}$ and s which are obtained from limited measurements. Small value of σ signifies that there is a high probability that the measurement will be near the mean value, i e. the distribution is peaked, while the large value of σ means a larger scatter. Fig. 3.3 illustrates Gaussian distributions along with the cumulative frequency distributions for small, medium and large values of σ respectively.

DETERMINATION OF $\bar{x}$ AND s FROM THE MEASUREMENTS

The mean and standard deviation can be calculated from the measurements by inserting various values into their respective formulae. However, considerable advantage is achieved by the use of scaling and shift of variable followed by the concept of assumed mean. Let there be a data set consisting of n measurements $x_1, x_2, \ldots x_r$ with their respective frequencies $f_1, f_2, \ldots\ldots\ldots f_r$.

A new variable $x_{Fi} = (x_i - c)/F_s$ is defined where c is a constant representing a shift of origin and F_s is the scale factor. By a proper choice of shift and scale factor, it may be possible to reduce the new variable which takes only the integral values. For this set of a new data set a mean m_0 is assumed, such that the new variable now is $x'_i = x_{Fi} - m_0$. The mean of the new distribution is $\bar{x}'$. It follows that

$$\bar{x}' = \bar{x}_F - m_0,$$

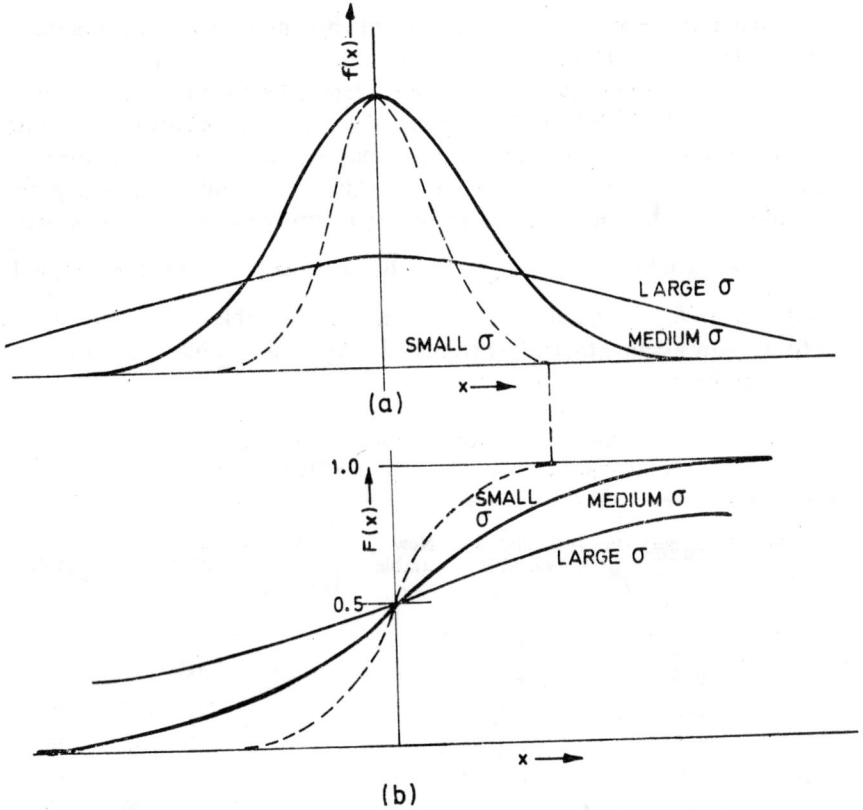

Fig. 3.3   (a) Gaussian distribution and (b) Its cumulative distribution
for different σ values

where $\bar{x}_F$ is the mean of x_{Fi} variables. It remains to be shown as to how $\bar{x}$ is connected with $\bar{x}'$ and similarly standard deviation s with s'.

It can be shown that

$$x_i = F_s x_{Fi} + c = F_s x_i' + F_s m_0 + c$$

Therefore the mean $\bar{x}$ is

$$\bar{x} = F_s \bar{x}' + F_s m_0 + c$$

In this equation but for $\bar{x}'$ which is obtained from the last data set, all other parameters are known and hence $\bar{x}$ is easily obtained. Similarly the variance s^2 is calculated from the last data set, i.e.

$$ns'^2 = \Sigma f_i (x_i' - \bar{x}')^2$$
$$= \Sigma f_i (x_{Fi} - \bar{x}_F)^2$$
$$= \frac{1}{F_s^2} \left[\Sigma f_i (x_i - \bar{x})^2 \right]$$

Thus $ns'^2 = \dfrac{1}{F_s^2} ns^2$ or $s = F_s s'$

The standard deviation s is then obtained by multiplying the standard deviation obtained from the reduced data set by the scale factor F_s.

In order to appreciate the advantage offered by this procedure consider the data of Table 3.1. First, group the original measurements into ranges instead of considering each measurement. This is known as Shepard's correction. Its effect on the value of $\bar{x}$ and s is negligibly small and may be ignored. Then follow the procedure described earlier, i.e., a new variable $x_F = \dfrac{x - 4.5}{10}$ is now generated from the original data. Here $F_s = 10$ and $c = 4.5$. The new variable thus runs from 0 to 9. Now assume that the mean is 5 i.e. $m_0 = 5$. Table 3.2 illustrates the procedure.

TABLE 3.2   Procedure to calculate $\bar{x}$ and s
(Assumed mean $m_0 = 5$, $F_s = 10$, $c = 4.5$)

S. No.	Original data x	Original data f	Lumped variable x	New variable x_F	$x'_i = (x_{Fi} - m_0)$	x'^2_i	fx'_i	fx'^2_i
1	0–9	1	4.5	0	−5	25	−5	25
2	10–19	3	14.5	1	−4	16	−12	48
3	20–29	4	24.5	2	−3	9	−12	36
4	30–39	11	34.5	3	−2	4	−22	44
5	40–49	20	44.5	4	−1	1	−20	20
6	50–59	29	54.5	5	0	0	0	0
7	60–69	22	64.5	6	1	1	22	22
8	70–79	8	74.5	7	2	4	16	32
9	80–89	1	84.5	8	3	9	3	9
10	90–99	1	94.5	9	4	16	4	16
Total		100					−26	252

From the table

$$\Sigma f_i (x_{Fi} - m_0) = \Sigma f_i x'_i = -26$$
$$\Sigma f_i (x_{Fi} - m_0)^2 = \Sigma f_i x'^2_i = 252.$$

Therefore

$$\bar{x}' = -\frac{26}{100} = -0.26,$$

and   hence $\bar{x} = 10 \times (-0.26) + 10 \times 5 + 4.5 = 51.9$

Also $$s'^2 = \frac{\Sigma f_i (x'_i - \bar{x}')^2}{n - 1} = \frac{\Sigma f_i x'^2_i - n(x')^2}{n - 1}$$

and
$$s = 10 \times \sqrt{\frac{252 - 100 \times 0.26 \times 0.\overline{26}}{100 - 1}}$$
$$= 15.74$$

Therefore the mean and standard deviation are 51.9 and 15.74 respectively.

CONFIDENCE LIMITS

It has been pointed out earlier that random errors obey the Gaussian distribution. When it is normalised such that the area under the curve is unity, it may be termed as Gaussian probability distribution function. The standard deviation is used as a measure of accidental error. If a large number of measurements are made, and these measurements suffer from random errors, it can be shown that

68.27% measurements lie within $\pm 1\sigma$ of $\bar{x}$

95.45% measurements lie within $\pm 2\sigma$ of $\bar{x}$, and

99.73% measurements lie within $\pm 3\sigma$ of $\bar{x}$.

The value $\pm a\sigma$ are called the confidence limits, where a (=1,2,3) depends on the chosen reliability. It is common to use a value of $a = 1.96$ which corresponds to a reliability of 95%. Another common method of stating bounds on the error uses the probable error e_p. This is defined as $e_p = 0.6745 s$. A range of $\pm e_p$ corresponds to 50% reliability. The standard devation σ is to be obtained from a very large (infinite) number of measurements. In practice the number of measurements taken naturally is small. Therefore the standard deviation σ must be calculated from the sample's standard deviation s. For this purpose conversion Table 3.3 and 3.4 may be used.

When the mean values of a physical quantity are obtained from large number of samples, the distribution of the mean values is also Gaussian with the mean μ and standard deviation $s/\sqrt{n}$, where n is the number of samples used to obtain mean values.

Conversion tables

(i) For the measurements

For 68.27% reliability $\bar{x} \pm 1\sigma = \bar{x} \pm t_1 s$

For 95% reliability $\bar{x} \pm 1.96\,\sigma = \bar{x} \pm t_2 s$

For 99.73% reliability $\bar{x} \pm 3\sigma = \bar{x} \pm t_3 s$

TABLE 3.3

n	2	5	10	20	50	100	∞
t_1	1.8	1.115	1.06	1.03	1.01	1.00	1.00
t_2	12.7	2.80	2.28	2.10	2.00	2.00	1.96
t_3	235	6.60	4.10	3.40	3.16	3.10	3.00

(ii)   For the mean $\bar{x}$

$$\text{For 68.27\% reliability } \bar{x} + \frac{1\sigma}{\sqrt{n}} = \bar{x} \pm \frac{t_1}{\sqrt{n}} s$$

$$\text{For 95\% reliability } \bar{x} + \frac{1.96\sigma}{\sqrt{n}} = \bar{x} \pm \frac{t_2}{\sqrt{n}} s$$

$$\text{For 99.73\% reliability } \bar{x} \pm \frac{3\sigma}{\sqrt{n}} = \bar{x} \pm \frac{t_3}{\sqrt{n}} s$$

TABLE 3.4

n	2	5	10	20	30	100	∞
$t_1/\sqrt{n}$	1.3	0.51	0.34	0.23	0.14	0.10	0
$t_2/\sqrt{n}$	9.0	1.24	0.72	0.47	0.28	0.20	0
$t_3/\sqrt{n}$	166	3.00	1.29	0.77	0.45	0.31	0

REJECTION OF MEASUREMENTS

It has been emphasised that $\pm 3\sigma$ confidence limits give a reliability that 99.73 % of all the measurements lie within these limits. Therefore, a measurement lying outside this range may be regarded as having a genuinely extraneous cause such as personal error in measurement, malfunction of the apparatus etc. and may therefore be rejected.

Testing a distribution for normalcy

The $\pm a\sigma$ bounds put on the best value of measurement are based on an assumption that the measurements obey Gaussian distribution. Therefore, it is advisable to test the data whether it follows the Gaussian distribution. Following two methods are available for this purpose

(i)   use of probability graph paper, and
(ii) X^2 test.

(i)   If one takes the cumulative distribution function for a Gaussian distribution and suitably distorts the vertical scale of the graph, the curve can be plotted as a straight line Such graph paper is commercially available and may be used to give a rough qualitative test for the conformity of the measurements to the Gaussian distribution.

(ii) Another method of testing for normalcy involves the use of X^2 (chi-square) statistical test. The X^2 test is applied by comparing the number of times n_0 an event was observed to occur with the number of times n_e that the event would be expected to happen if the hypothesis were true. X^2 is then defined as

$$X^2 = \sum \frac{(n_0 - n_e)^2}{n_e}$$

The hypothesis is that the measurements follow the Gaussian distribution. The value of n_e can be obtained from the known values of n, $\bar{x}$ and s using the Gaussian distribution formula and conversion tables. The calculation of the X^2 for the measurements of Table 3.1 has been done and presented in Table 3.5. The only point that needs consideration is to group together thinly populated regions at both the ends of the distribution.

TABLE 3.5 Procedure to calculate X^2
($\bar{x} = 4.74$, $s = 1.57$, $n = 100$)

S. No.	x_F	n_0	n_e	$(n_0 - n_e)^2$	$(n_0 - n_e)^2/n_e$
1	0−1	4	1.5	6.25	4.17
2	2	4	5.57	2.46	0.44
3	3	11	13.76	7.62	0.55
4	4	20	22.69	7.24	0.32
5	5	29	25.00	16.00	0.64
6	6	22	18.40	12.96	0.70
7	7	8	9.04	1.08	0.12
8	8−9	2	2.97	0.98	0.32

Total $= 7.26 = \chi^2$

In order to apply the criterion, one finds the degrees of freedom. There are eight rows in Table 3.5, and three constants n, $\bar{x}$ and s that are used to fit the data. Therefore, the degrees of freedom are 5. For the degrees of freedom and experimental value of χ^2, the Table 3.6 gives the probability P of the chance occurrence of this value of X^2 or higher values. If $X^2 = 0$, the expected and the experimental distributions match exactly, thus giving very high values of P. However, one must be very cautious in making inferences with high values of P. As an example, consider a float level controller used to control the level of water, where the recorder always shows the set point level with no deviations, whatsoever. This is not experimentally true as the controller does not work that way and the experimenter will be suspicious to look for faults in the recorder. On the other hand, larger the value of X^2, the larger the disagreement between the expected and the experimental distributions, or smaller is the probability that the observed and expected distributions match with each other. A good rule of thumb is that if P lies between 0.1 and 0.9, the observed distribution may be considered to follow the expected distribution.   For the values of P below 0.02 and higher than 0.98, the expected distribution may be considered unlikely.

It is worthwhile to point out one feature of χ^2 that has general applications.   Inspection of Table 3.6 will reveal that if the hypothesis is true, χ^2 is of the order of magnitude of the number of degrees of freedom.   Therefore, each individual term $\dfrac{(n_0 - n_e)^2}{n_e}$ will be of the order of unity, i.e.

$$(n_0 - n_e)^2/n_e \simeq 1$$

$$\text{or } n_0 \approx n_e + \sqrt{n_e}.$$

The interpretation of this result is that whenever the value of n_e is to be obtained it is likely that $n_e \pm \sqrt{n_e}$ is observed.   For example, if a coin is tossed 1000 times, one expects 500 heads but in practice it is likely to be off by a number of the order of magnitude of $\sqrt{500}$ or 22. As the number of measurement increases, the total error increases as $\sqrt{n_e}$, while the fractional error decreases as $n_e^{-1/2}$.

METHOD OF LEAST SQUARES–LINE OF BEST FIT

For most of the instruments, but not all, the input-output relation is ideally linear.   In some of the instruments where input-output relation is non-linear, the operation is restricted over a linear range. The average calibration curve for such an instrument is generally taken as a straight line that fits the scattered data points best.   The most common criterion for the best line fit is the least-squares method. The equation of a straight line is taken as

$$q_0 = mq_i + a,$$

where q_0 and q_i are the output and input quantities respectively, and m and a are the slope and the intercept of the straight line. According to the method of least-squares, the sum of the squares of the vertical deviations of the data points from the fitted line should be minimum in order to obtain the value of m and a.   Therefore, one minimises a quantity 'S' given by

$$S = \overset{n}{\Sigma}[q_0 - (mq_i + a)]^2$$

by setting $\dfrac{\partial S}{\partial m} = \dfrac{\partial S}{\partial a} = 0$. This gives a pair of equations in m and a which can be solved for their values.   These values are given by

$$m = \frac{n\Sigma q_i q_0 - \Sigma q_i \Sigma q_0}{n\Sigma q_i^2 - (\Sigma q_i)^2},$$

and

$$a = \frac{\Sigma q_0 \Sigma q_i^2 - \Sigma q_i q_0 \Sigma q_i}{n\Sigma q_i^2 - (\Sigma q_i)^2}.$$

where n is the total number of data points.   Assuming that q_i were fixed, and repeated measurements were made on q_0, the latter will exhibit scatter

TABLE 3.6 Chi-squared—[P is the probability that the value in the table will be exceeded for a given number of degrees of freedom F]

$F=$	0.995	0.990	0.975	0.950	0.900	0.750	0.500	0.250	0.100	0.050	0.025	0.010	0.005
1	0.0^4393	0.0^3157	0.0^3982	0.0^2393	0.0158	0.102	0.455	1.32	2.71	3.84	5.02	6.63	7.88
2	0.0100	0.0201	0.0506	0.103	0.211	0.575	1.39	2.77	4.61	5.99	7.38	9.21	10.6
3	0.0717	0.115	0.216	0.352	0.584	1.21	2.37	4.11	6.25	7.81	9.35	11.3	12.8
4	0.207	0.297	0.484	0.711	1.06	1.92	3.36	5.39	7.78	9.49	11.1	13.3	14.9
5	0.412	0.554	0.831	1.15	1.61	2.67	4.35	6.63	9.24	11.1	12.8	15.1	16.7
6	0.676	0.872	1.24	1.64	2.20	3.45	5.35	7.84	10.6	12.6	14.4	16.8	18.5
7	0.989	1.24	1.69	2.17	2.83	4.25	6.35	9.04	12.0	14.1	16.0	18.5	20.3
8	1.34	1.65	2.18	2.73	3.49	5.07	7.34	10.2	13.4	15.5	17.5	20.1	22.0
9	1.73	2.09	2.70	3.33	4.17	5.90	8.34	11.4	14.7	16.9	19.0	21.7	23.6
10	2.16	2.56	3.25	3.94	4.87	6.74	9.34	12.5	16.0	18.3	20.5	23.2	25.2
11	2.60	3.05	3.82	4.57	5.58	7.58	10.3	13.7	17.3	19.7	21.9	24.7	26.8
12	3.07	3.57	4.40	5.23	6.30	8.44	11.3	14.8	18.5	21.0	23.3	26.2	28.3
13	3.57	4.11	5.01	5.89	7.04	9.30	12.3	16.0	19.8	22.4	24.7	27.7	29.8
14	4.07	4.66	5.63	6.57	7.79	10.2	13.3	17.1	21.1	23.7	26.1	29.1	31.3
15	4.60	5.23	6.26	7.26	8.55	11.0	14.3	18.2	22.3	25.0	27.5	30.6	32.8

(*Contd.*)

P / F=	0.995	0.990	0.975	0.950	0.900	0.750	0.500	0.250	0.100	0.050	0.025	0.010	0.005
16	5.14	5.81	6.91	7.96	9.31	11.9	15.3	19.4	23.5	26.3	28.8	32.0	34.3
17	5.70	6.41	7.56	8.67	10.1	12.8	16.3	20.5	24.8	27.6	30.2	33.4	35.7
18	6.26	7.01	8.23	9.39	10.9	13.7	17.3	21.6	26.0	28.9	31.5	34.8	37.2
19	6.84	7.63	8.91	10.1	11.7	14.6	18.3	22.7	27.2	30.1	32.9	36.2	38.6
20	7.43	8.26	9.59	10.9	12.4	15.5	19.3	23.8	28.4	31.4	34.2	37.6	40.0
21	8.03	8.90	10.3	11.6	13.2	16.3	20.3	24.9	29.6	32.7	35.5	38.9	41.4
22	8.64	9.54	11.0	12.3	14.0	17.2	21.3	26.0	30.8	33.9	36.8	40.3	42.8
23	9.26	10.2	11.7	13.1	14.8	18.1	22.3	27.1	32.0	35.2	38.1	41.6	44.2
24	9.89	10.9	12.4	13.8	15.7	19.0	23.3	28.2	33.2	36.4	39.4	43.0	45.6
25	10.5	11.5	13.1	14.6	16.5	19.9	24.3	29.3	34.4	37.7	40.6	44.3	46.9
26	11.2	12.2	13.8	15.4	17.3	20.8	25.3	30.4	35.6	38.9	41.9	45.6	48.3
27	11.8	12.9	14.6	16.2	18.1	21.7	26.3	31.5	36.7	40.1	43.2	47.0	49.6
28	12.5	13.6	15.3	16.9	18.9	22.7	27.3	32.6	37.9	41.3	44.5	48.3	51.0
29	13.1	14.3	16.0	17.7	19.8	23.6	28.3	33.7	39.1	42.6	45.7	49.6	52.3
30	13.8	15.0	16.8	18.5	20.6	24.5	29.3	34.8	40.3	43.8	47.0	50.9	53.7

From Catherine M. Thomson, Biometrika, **32**, 1941 as abridged by Harold J. Larsow "Introduction to Probability Theory and Statistical Inference" 2nd Ed., John Wiley & Sons, New York, 1974.

for any fixed value of q_i. Therefore m and a will also exhibit scatter. It can be shown that the standard deviations s_m and s_a for m and a respectively are given by the following expressions

$$s_m^2 = \frac{n s_{q0}^2}{n \Sigma q_i^2 - (\Sigma q_i)^2},$$

and

$$s_a^2 = \frac{s_{q0}^2 \Sigma q_i^2}{n \Sigma q_i^2 - (\Sigma q_i)^2},$$

where

$$s_{q0}^2 = \frac{\Sigma [q_0 - (m q_i + a)]^2}{n} \text{ is the variance of } q_0.$$

Assuming that the measurements follow Gaussian distribution and 95% reliability limits, one can set the bounds on both m and a as

$$m = m \pm 1.96 \, s_m,$$

$$\text{and} \quad a = a \pm 1.96 \, s_a.$$

This procedure is used to obtain the calibration curve where observations at constant input are made and a best line fit is accomplished. This linear calibration curve is later used to obtain the value of input quantity from the output value   It is, therefore, necessary to put bounds on the input quantity which is being measured. From the line obtained by the method of least-squares,

$$q_i = \frac{q_0 - a}{m},$$

The q_i values computed this way must have some errors. The standard deviation s_{q_i} of q_i is computed from

$$s_{qi}^2 = \frac{1}{n} \Sigma \left(\frac{q_0 - a}{m} - q_i \right)^2 = \frac{s_{q_0}^2}{m^2}.$$

Thus the 95% reliability limits on q_i are $q_i = q_i \pm 1.96 \left(\frac{s_{q_0}}{m} \right)$. It must be noted that if s_{q_0} is assumed to be same for any input value q_i, a set of data q_0 need not be obtained for the same value of q_i, rather the standard deviation s_{q_0} can be obtained from the set of data where an output q_0 is measured for any input value q_i.

EXAMPLE: Following data points are expected to follow a linear relationship $q_0 = m q_i + a$. Obtain the values of m and a using a method of least squares. Also calculate the standard deviation of m and a respectively.

q_i (units)	0	1	2	3	4	5	6	7	8	9	10
q_0 (units)	-1.01	0.25	1.09	2.34	3.62	4.71	5.60	6.83	7.90	8.89	9.93

Solution: The calculation is best done by a suitable table given here (Table 3.7).

TABLE 3.7

q_i	q_0	q_i^2	$q_0 q_i$	q_{0c}	$(q_0 - q_{0c})$	$(q_0 - q_{0c})$
0	−1.01	0	0	−0.931	−0.079	0.006
1	0.25	1	0.25	0.167	0.083	0.007
2	1.09	4	2.18	1.265	−0.175	0.031
3	2.34	9	7.02	2.363	−0.023	0.001
4	3.62	16	14.48	3.461	0.159	0.025
5	4.71	25	23.55	4.559	0.151	0.023
6	5.60	36	33.60	5.657	0.057	0.003
7	6.83	49	47.81	6.755	0.075	0.006
8	7.90	64	63.20	7.853	0.047	0.002
9	8.89	81	80.01	8.951	−0 067	0.004
10	9.93	100	99.30	10.049	−0.119	0.014
55	50.15	385	371.4			0.121

From the equations for m and a.

$$m = \frac{n\Sigma q_0 q_i - \Sigma q_i \, \Sigma q_0}{n\Sigma q_i^2 - (\Sigma q_i)^2},$$

$$= \frac{11 \times 371.4 - 55 \times 50.15}{11 \times 385 - 55 \times 55} = 1.098$$

$$a = \frac{\Sigma q_0 \Sigma q_i^2 - \Sigma q_0 q_i \Sigma q_i}{n\Sigma q_i^2 - (\Sigma q_i)^2}$$

$$= \frac{50.15 \times 385 - 371.4 \times 55}{11 \times 385 - 55 \times 55} = -0.931 \text{ units}$$

Thus $q_0 = 1.098 \, q_i - 0.931$.

In order to calculate the standard deviations for m and a, the standard deviation s_{q_0} for q_0 is first calculated. Let the values of q_0 obtained from best fit line be q_{0c}. The standard deviation s_{q_0} is thus obtained from

$$s_{q_0}^2 = \frac{\Sigma (q_0 - q_{0c})^2}{n} = \frac{0.121}{11} = 0.011$$

Thus $s_{q_0} = 0.105$ units.

Now $\quad s_m^2 = \dfrac{11 \times 0.011}{11 \times 385 - 55 \times 55} = 1 \times 10^{-4}$.

or $\quad s_m = 1 \times 10^{-2}$ units

Similarly, $\quad s_a^2 = \dfrac{0.105 \times 0.105 \times 385}{11 \times 385 - 55 \times 55} = 3.51 \times 10^{-3}$

or $\quad s_a = 5.92 \times 10^{-2}$ units

The result may, therefore, be expressed for 95% reliability, as

$$m = 1.098 \pm 1.96 \times 10^{-2} \text{ and}$$

$$a = -0.931 \pm 0.116 \text{ Units.}$$

PROPAGATION OF ERROR

The nomenclature followed here is:

$$\text{reading} \pm \text{uncertainty} = \text{result}$$

The term 'reading' should be called 'corrected reading' after applying the correction. The correction has the same magnitude as the systematic error, but opposite sign. It is, however, a usual practice to call the 'corrected reading' as a 'reading' only.

In many experiments, results from different measuring instruments are used to compute the value of a particular physical quantity. The measuring instruments will in general have different uncertainties of measurement, and hence the value of the physical quantity may also be uncertain to a certain extent. The following two questions now need some consideration:

(i) If the uncertainty of each instrument is known, what is the uncertainty of the computed result?

(ii) If a certain uncertainty in the computed result is desired, what uncertainty is allowed in the measurement by each instrument?

These questions can be answered by considering a problem of computing a quantity, g. The quantity g is a known-function of n independent variables $u_1, u_2 \ldots, u_n$. It is expressed in a functional form as

$$g = f(u_1, u_2 \ldots, u_n).$$

The variables $u_1, u_2, \ldots, u_n$ are quantities measured with different instruments. Let these values be uncertain by $\pm \Delta u_1, \pm \Delta u_2, \ldots, \pm \Delta u_n$. These uncertainties will result in an uncertainty of magnitude $\pm \Delta g$ in g. The method of computation of Δg depends on the following two cases:

(a) when $\pm \Delta u_n$ is an absolute limit on the uncertainties as is the case with systematic errors, and

(b) when $\pm \Delta u_n$ is a statistical bound such as $\pm 1.96s$ or e_p limits as is the case with systematic errors of unknown sign or accidental errors.

Case (a): $g = f(u_1, u_2, \ldots, u_n).$

Therefore

$$g \pm \Delta g = f(u_1 \pm \Delta u_1, u_2 \pm \Delta u_2 \ldots, u_n \pm \Delta u_n)$$

If $\Delta u_n \ll u_n$, the right hand side of the above equation can be expanded in the Taylor's series. Thus

$$g \pm \Delta g = f(u_1, u_2 \ldots u_n) \pm \Delta u_1 \frac{\partial f}{\partial u_1} \pm \Delta u_2 \frac{\partial f}{\partial u_2} + \ldots + \ldots \pm \Delta u_n \frac{\partial f}{\partial u_n} + \text{higher}$$

order terms.

Therefore the absolute error E_a is

$$E_a = \left| \Delta g \right| = \left| \Delta u_1 \frac{\partial f}{\partial u_1} \right| + \left| \Delta u_2 \frac{\partial f}{\partial u_2} \right| + \ldots + \left| \Delta u_n \frac{\partial f}{\partial u_n} \right|,$$

where absolute values have been used; the total error is, therefore, obtained by adding the numerical values of individual errors. This form of equation is very useful since it shows which variable (u_n) exerts the strongest influence on the uncertainty of overall result.

The relative percentage error E_r is given by

$$E_r = \frac{|\Delta g|}{g} \times 100 = 100 \frac{E_a}{g}.$$

The computed result may be expressed as either $g + E_a$ or $g + E_r$ and the interpretation is that this error will not be exceeded in the measurements since Δu_n is defined this way.

When the second question is considered, i.e. where a certain overall uncertainty in the result is desired and the uncertainty of measurement by each measuring instrument is to be calculated, the problem becomes extremely complicated. It is simplified when the 'method of equal effects' is used, i.e. it is assumed that each instrument measures such that

$$\left| \Delta u_1 \frac{\partial f}{\partial u_1} \right| = \left| \Delta u_2 \frac{\partial f}{\partial u_2} \right| = \left| \Delta u_3 \frac{\partial f}{\partial u_3} \right| = \ldots = \left| \Delta u_n \frac{\partial f}{\partial u_n} \right|.$$

Or

$$n\Delta u_i \frac{\partial f}{\partial u_i} = \Delta g \, (i = 1, 2, 3, \ldots n)$$

and hence

$$\Delta u_i = \frac{\Delta g}{n(\partial f / \partial u_i)}.$$

This gives the uncertainty of measurement of each instrument. In actual practice it may turn out that some instruments cannot meet the accuracy demand imposed by this condition. However if other instruments can measure to accuracies smaller than Δu_i it may still be possible to meet overall uncertainty requirements. The experimenter should carefully choose various instrument systems.

Case (b): When Δu_n is considered as statistical bound such as $\pm 1.96s$ limits i.e. measurements from each instrument follow the normal distribution, the computed result will also follow the Gaussian distribution. Thus, the proper method to obtain the uncertainty in the computed result is to properly combine the measurement uncertainties by root-sum-square formula i.e.

$$E_{a_{rss}} = \left[\left(\Delta u_1 \frac{\partial f}{\partial u_1}\right)^2 + \left(\Delta u_2 \frac{\partial f}{\partial u_2}\right)^2 + \ldots + \left(\Delta u_n \frac{\partial f}{\partial u_n}\right)^2\right]^{1/2}$$

The computed result can be then expressed as

$$g \pm a E_{a_{rss}},$$

where a takes values 1, 2 or 3 depending on the reliability of the result. The above equation always gives a smaller value of error than given by the equation under case (a).

Further, when a known uncertainty in the computed result is desired, the problem is solved only using the method of equal effects. The uncertainty of each instrument is given by

$$\Delta u_i = \frac{E_{a_{rss}}}{\sqrt{n}\,(\partial f/\partial u_i)} : (i = 1, 2 \ldots n)$$

These statistical concepts developed thus far should be implemented in regular experimentation.

Exercises

1. It is desired to have 150Ω resistance arrangement either using two $75\Omega \pm 0.03\,\Omega$ resistances in series or two $300\Omega \pm 0.15\Omega$ resistances in parallel. Which of the two configurations should be used? Calculate the uncertainty for each configuration.

2. The discharge coefficient C of an orifice is given by

$$C = \frac{W}{t\rho A \sqrt{2gh}}$$

Calculate the value of C with its uncertainty for the following data

$$W = (400 \pm 0.2) \text{ Kg}, \; A = \pi d^2/4, \; d = (15 \pm 0.02) \text{ mm}$$
$$t = (500 \pm 2) \text{ sec}, \; g = (9.81 \pm 0.1\%) \text{ m/sec}^2$$
$$\rho = (10^3 \pm 0.1\%) \text{ Kg/m}^3, \; h = (4 \pm 0.05) \text{ m}.$$

The uncertainty of C may be calculated for the following cases:
(a) the uncertainties are absolute values, and
(b) the uncertainties are $\pm 3s$ statistical limits.

3. In a strain gauge experiment, the output e_0 is related to the strain ϵ through the relation

$$\epsilon = \frac{e_0}{V} \frac{1}{G_F} \frac{(R_0 + R_s)^2}{R_0 \cdot R_s}$$

Calculate the uncertainty in ϵ when the output voltage e_0, excitation voltage V, gauge factor G_F, resistances R_0 and R_s are measured to $\pm 2\%$.

4. The frequency of a variable x is given by $f = Ce^{-x/a}$ for all positive values of x. C and a are constants. Find the mean and standard deviation of the distribution.

5. A sample of 400 data has a mean of 2.5 units. Test whether it can be regarded as a random sample drawn from a normal distribution with mean of 2.6 units and standard deviation of 0.5 units.

6. The following functional relations are given

(1) $y = ax^b$, (2) $y = ae^{bx}$, (3) $y = a + bx + cx^2$, (4) $y = \dfrac{x}{a+bx} + c$.

Suggest suitable modifications so that these relations can be plotted as straight lines.

7. For the following set of data points, q_0 is expected to be a quadratic function of q_i. Obtain the quadratic function by the method of least squares.

q_i	1	2	3	4	5	6
q_0	2.1	8.9	19.3	40.0	80.0	110.0

8. The expected relationship in a measurement between q_0 and q_i is of the form

$$q_0 = bq_i^a .$$

Obtain the best fit by the method of least squares, given,

q_i	1720	1350	1600	1000	1240	2150	2470	2600	2030
q_0	13.0	15.0	19.0	18.0	19.2	21.5	28.7	32.1	33.0

q_i	2980	3400	3000	3300	2100	3900
q_0	42.7	44.0	52.0	60.0	80.0	130.0

Measurement of Force and Torque

4.1 Introduction

Force is represented mathematically as a vector and has a point of application. Therefore the measurement of force involves the determination of its magnitude as well as its direction. The measurement of force may be done by any of the two methods:

(i) *Direct methods:* These involve a direct comparison with a known gravitational force on a standard mass, say by a balance.

(ii) *Indirect methods:* These involve measurement of the effect of force on a body, for example:

(a) measurement of acceleration of a body of a known mass that is subjected to force, and

(b) measurement of resultant effect (deformation) when the force is applied to an elastic member.

4.2 Direct methods

A body of mass m in the earth's gravitational field experiences a force that is given by

$$W = mg,$$

where W is the weight of the body. Any unknown force may be compared with the gravitational force (mg) on the standard mass m. The values of m and g, the acceleration due to gravity, should be known accurately in order to know the magnitude of the gravitational force.

The mass is a fundamental quantity, and its standard, kilogram, is kept in a vault at Sevres, France. The other masses can be compared with this standard with a precision of a few parts in 10^9. On the other hand, g is a derived quantity but still makes a convenient standard. Its value can be measured with an accuracy of 1 part in 10^6. Therefore an unknown force can be compared with the gravitational force with an accuracy of about this order of magnitude.

ANALYTICAL BALANCE

Direct comparison of an unkown force with the gravitational force can be illustrated with the help of an analytical balance. The direction of force is parallel to that of the gravitational force, and hence only its magnitude needs to be determined. The constructional details of an analytical balance shown schematically in Fig. 4.1, can be found some-where else. Suffice to say that the balance arm rotates about the knife edge at point O, and the two forces W_1 and W_2 are applied at the ends of the arm. W_1 is an unknown force and W_2 is the known force due to a standard mass. Point G is the centre of gravity of the arm, and W_B is the weight of the balance arm and the pointer acting at G. Fig. 4.1 shows the balance arm in an unbalanced position when the force W_1 and W_2 are unequal. This unbalance is indicated by the angle θ which the pointer makes with the vertical. In the balanced position $W_1 = W_2$, and hence θ is zero. Therefore, the weight of the balance arm and the pointer do not influence the measurements.

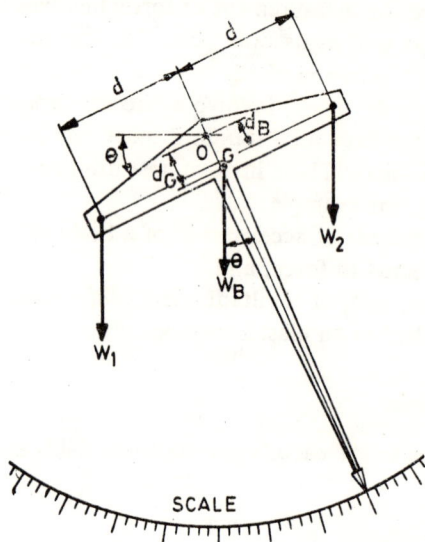

Fig. 4.1   Schematic of an analytical balance

Sensitivity S of the balance is defined as the angular deflection per unit unbalance between the two weights W_1 and W_2. It is given by

$$S = \frac{\theta}{W_1 - W_2} = \frac{\theta}{\Delta W},$$

where ΔW is the difference between W_1 and W_2. The sensitivity S can be calculated by writing moment equation at equilibrium i.e.

$$W_1(d \cos \theta - d_B \sin \theta) = W_2(d \cos \theta + d_B \sin \theta) + W_B d_G \sin \theta,$$

where the distances d_B, d_G and d are shown in Fig. 4.1. Assuming small deflection, the above equation is rewritten as

$$W_1(d - d_B \theta) = W_2(d + d_B \theta) + W_B d_G \theta.$$

Therefore the sensitivity S can be expressed as

$$S = \frac{\theta}{W_1 - W_2} = \frac{d}{(W_1 + W_2)d_B + d_G W_B}.$$

At a near balance point $W_1 \simeq W_2 = W$ and hence

$$S = \frac{\theta}{\Delta W} = \frac{d}{2W d_B + W_B d_G}.$$

The sensitivity of the balance will be independent of the weight W provided it is so designed that $d_B = 0$, then

$$S = \frac{d}{W_B d_G}.$$

The sensitivity depends on the construction parameters of the balance arm and is independent of the weights being compared. That is a very important result. The sensitivity can be improved by decreasing both d_G and W_B and increasing d. A compromise, however, is to be struck between the sensitivity and stability of the balance. Precision balances are available which have an accuracy of 1 part in 10^9.

While comparing an unknown force with a gravitational force acting on a known mass using a balance, sufficient care should be exercised to correct the result for buoyancy forces acting on the mass(es). The range of measurement can be considerably extended by using levers as is done in weighing platforms.

4.3 Indirect methods

ACCELERATION MEASUREMENT

The measurement of force by measuring acceleration of a standard mass M when force is acting on this is based on the principle that

$$F = Ma, \text{ where '}a\text{' is the acceleration.}$$

The measurement of 'a' can be carried out by accelerometers. However, this method is of limited application since the force determined is the resultant force acting on the mass. Often, several unknown forces are acting and they cannot be separately measured by this method. This method is represented schematically in Fig. 4.2.

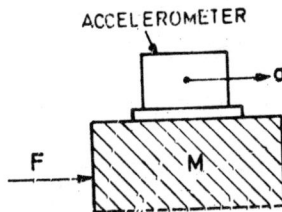

Fig. 4.2 Force mesurement by accelerometer

MEASUREMENT WITH ELASTIC ELEMENTS

Elastic elements are frequently used for the measurement of force because of their large range, continuous monitoring, ease of operation and ruggedness. They furnish an indication of the magnitude of force through displacement measurement. They are used both for dynamic and static force measurements. A few of them are described here.

Springs

A simple spring is an example of a force displacement transducer. If the displacement from the equilibrium position is y, then the force F is given by (within a certain limit)

$$F = Ky$$

where K, force per unit displacement, is the spring constant. The displacement bears a linear relation with the applied force. This system is quite satisfactory for steady state measurements.

Axially Loaded Members

In this method the members (say, a bar or a rod) are axially loaded either in compression or tension. The displacement y of the free end, when a force F is applied, is given by

$$y = \frac{FL}{AE}$$

where A is the area, $L =$ length, and $E =$ Young's modulus of the material. The spring oonstant of the member may be represented by $K = AE/L$. Therefore $F = Ky$.

Figure 4.3 (a), (b), (c) show three different arrangements for measuring displacement y. In the first arrangement, the displacement is

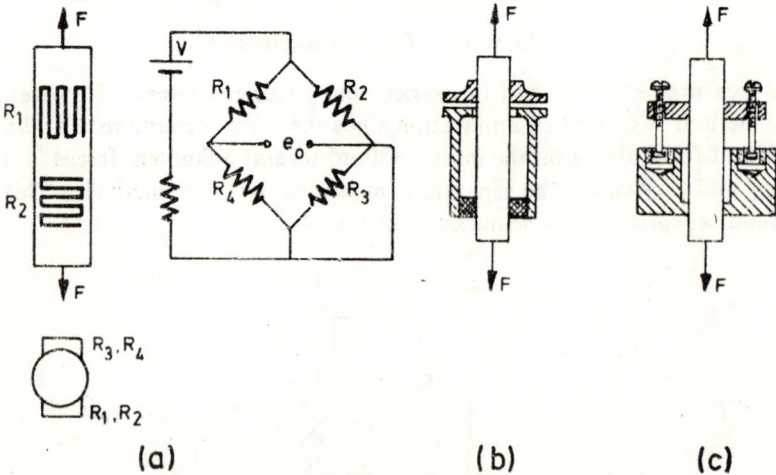

(a)   (b)   (c)

Fig. 4.3   Axially loaded members for force measurement

measured by measuring the change in resistances of the strain-gauges, while in the second and third arrangements capacitance and inductance changes are measured. The inductance gauges are linear and are obviously not effected by the bending of the member. On the other hand, capacitance gauges are non-linear, and will be influenced, to a certain extent, by the bending of the member.

For measurement of displacement, the strain-gauges are mounted as shown in Fig. 4.3(a). The specimen is subjected to a force, and the strain is measured by the gauges. The gauges R_1, R_3 measure the axial strain (y/L), while the gauges R_2, R_4 measure the lateral strain. This particular arrangement of mounting the gauges is such that the influences of bending and temperature are compensated. The compensation is almost complete if the gauges are mounted symmetrically and the gauge characteristics are sufficiently uniform. For the gauges shown in figure, the strains are given by

$$\epsilon_1,\ \epsilon_3 = F/AE,$$

and
$$\epsilon_2,\ \epsilon_4 = -\ \mu F/AE,$$

where μ is the Poisson's ratio. The bridge output e_o is

$$e_0 = \frac{VG_F}{4}(\epsilon_1 - \epsilon_2 + \epsilon_3 - \epsilon_4) = \frac{VG_F}{2}(1 + \mu)\frac{F}{AE}.$$

The bridge output thus provides the value of F/AE.

Cantilever beams

Cantilevers are ideally suited for the measurement of bending moments about two axes mutually perpendicular to the axis of the beam. If the point of application of load along the beam is known, the bending moment can be directly translated into force. Fig. 4.4 represents

Fig. 4.4   Cantilever elastic element

a cantilever, where a force F is applied at a distance L from the fixed end. The deflection of the free end is related to the force by the following equation

$$F = \frac{3EIy}{L^3}$$

where I is the moment of inertia of the cross section of the beam. For a cantilever of rectangular cross section of width b and thickness h, the moment of inertia $I = \dfrac{bh^3}{12}$. Therefore the deflection y is

$$y = \frac{4FL^3}{Ebh^3}.$$

The spring constant of the system is therefore given by

$$K = \frac{Ebh^3}{4L^3}.$$

The strains induced by the application of force are measured by the strain gauges. The strain gauges R_1, R_2, R_3 and R_4 are used to measure the force F while R'_1, R'_2, R'_3 and R'_4 are used to measure the force P. The strain gauges are symmetrically mounted with respect to the neutral axis of the beam and are wired as shown in Fig. 4.5. The arrangement is such that the bridge acts as a full bridge and the influence of temperature is compensated. The strain is obtained from the equations:

$$\frac{M_x}{I} = \frac{\sigma_x}{c},$$

$$\epsilon = \frac{\sigma_x}{E}$$

where $c = h/2$, $M_x = Fl$, l is the distance from the center of gauges to the point of application of the force. This gives

$$\epsilon = \frac{6Fl}{Ebh^2}.$$

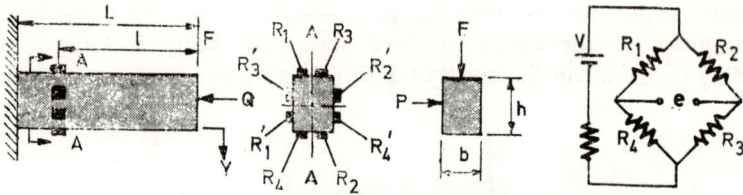

Fig. 4.5   Cantilever elastic element for measuring force

The strain gauges R_1, R_3 measure the tensile strain while R_2, R_4 measure the compressive strains. The strains ϵ_1, ϵ_2, ϵ_3 and ϵ_4 as measured by the gauges are of equal magnitude of $(6\ Fl/Ebh^2)$. Similar equations hold good for the other component of force. It may be noted that an axial force component Q will not affect the measurements provided it acts at the center of cross section.

Rings

One of the very useful and important devices under this heading is a proving ring. This has been the standard for calibrating tensile-testing machines and is, in general, the means whereby accurate measurement of large static loads may be made. Proving ring can be used over a wide range of loads starting from 1500 to 15×10^5N.

Figure 4.6 shows a compression-type proving ring. The compressive load deforms the ring; a sensitive micrometer is employed for the deflection measurement. Bosses are provided to clamp the ring rigidly to avoid rotation. To obtain a precise measurement, one edge of the micrometer is mounted on a vibrating reed device which is plucked to obtain a vibratory motion. The micrometer contact is moved forward until a noticeable damping of the vibration is observed. Deflection measurements may be made within $\pm$ 0.50 μm with this method.

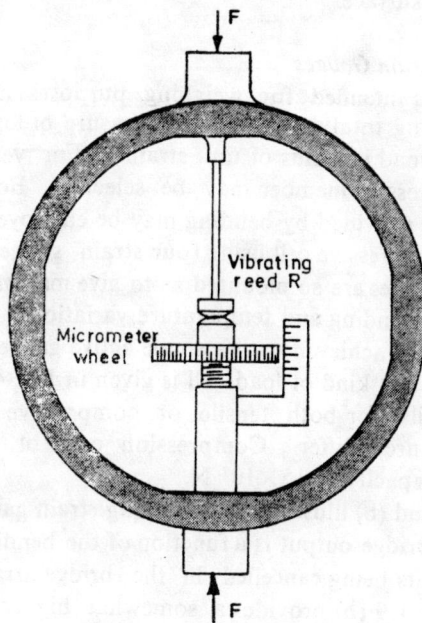

Fig. 4.6 Proving ring

In another variant of proving ring, a differential transformer is used for the measurement of deflection.

Another elastic device which is frequently used for force measurement is a thin ring shown in Fig 4.7. The force F and deflection y relation for a thin ring is

$$F = \frac{16}{\pi/2 - 4/\pi} \frac{EI}{d^3} y$$

where d is the outside ring diameter and I is the moment of inertia about the centroidal axis of the bending section. This equation is derived under the assumption that the thickness of the ring is small compared with the radius. The force is applied perpendicular to the axis of the ring as shown in Fig 4.7.

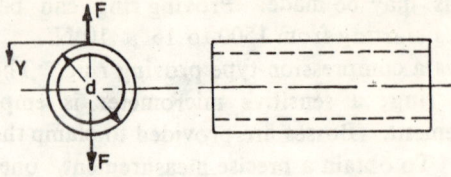

Fig. 4.7 Thin ring elastic element

Other types of rings used for force measurement are octagonal rings and extended octagonal rings. They are designed so as to have no rotation of the top surface.

Load Cells Using Strain Gauges

Force transducers intended for weighing purposes are called load cells. Instead of using total deflection as a measure of load, strain-gauge load cells measure load in terms of unit strains. For very large loads, direct tensile-compressive member may be selected. For small loads, strain amplification provided by bending may be employed to advantage. One such tensile compressive cell using four strain gauges is shown in Fig 4.3.(a). The gauges are so mounted as to give maximum output, and compensation for bending and temperature variations. The sensitivity is $2 (1 + \mu)$ times that achieved with a single active gauge in the bridge. Another version of this kind of load cell is given in Fig. 4.8. The basic arrangement is similar for both tensile or compressive load measurements, only the fixtures differ. Compression cells of this kind have been used with a capacity of 15×10^6 N.

Figures 4.9 (a) and (b) illustrate proving ring strain gauge load cells. In Fig 4.9 (a) the bridge output is a function of the bending strains only, the axial components being cancelled in the bridge arrangement. The arrangement of Fig 4.9 (b) provides a somewhat higher sensitivity because the output includes both the bending and axial components sensed by gauges R_1 and R_4.

Temperature sensitivity of load cells using strain gauges: So far relatively ideal situations have been considered where the compensation both for bending and temperature variations was achieved by a particular mounting arrangement of the gauges However, careful analysis indicates that the temperature influences the measurements in the following two ways:

    (i)   change of dimensions, and

    (ii) change in the Young's modulus of the material.

Of these, the latter influences the measurements more. In a strain-gauge bridge circuitary, the bridge sensitivity is made to vary with temperature in a direction opposite to that caused by the variation of Young's modulus of the material. As an example, the Young's modulus of a material of a load cell, usually, decreases with temperature, and hence its deflection increases; the material becomes more springy and hence deflects more for a given load. This results in an increase in the sensitivity of the load cell. This increased sensitivity is off set by reducing the sensitivity of the bridge by the use of a thermally sensitive compensating resistance R_s as shown in Fig. 4.10 (a). In order to carry out calibration, usually two resistances each of nominal value of $(R_s/2)$ are used as shown in Fig 4.10 (b). In order to calculate the value of R_s required to compensate for the temperature influence due to (ii), an analysis of the bridge-circuit is required. It can be shown that the

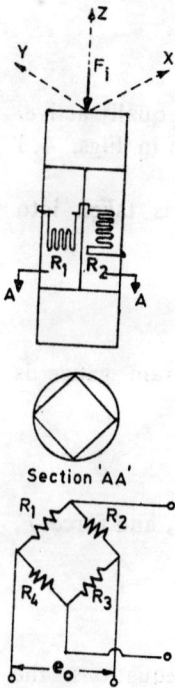

Fig. 4.8 Strain
gauge load cell

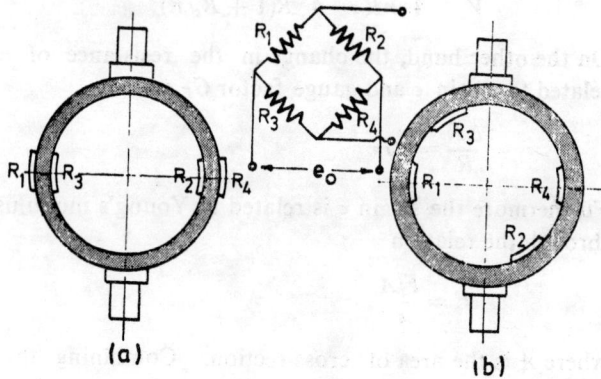

Fig. 4.9 Strain gauge cell

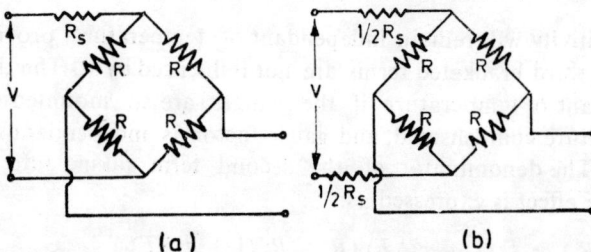

Fig. 4.10 Strain gauge bridge with compensation resistance

introduction of a resistance, R_s, in an input lead reduces the sensitivity of an equal-arm bridge, each arm having a strain gauge of nominal resistance R, by a factor n. The factor n, also called the bridge factor, is given by

$$n = \frac{1}{1 + (R_s/R)}$$

Further from the bridge analysis, it can be proved that the potential difference Δe across an indicating instrument, provided the bridge is initially balanced, due to a change ΔR in any resistance R is

$$\frac{\Delta e}{V} = \frac{m}{4} \frac{\Delta R}{R},$$

where $m = 4$ for a full-bridge in which all four gauges are equally active. In examples of tensile and compression load cells as shown in Figs. 4.3 and 4.8, the value of m is $2(1 + \mu)$.

When the effect of bridge factor due to resistance R_s is taken into account, the above equation becomes

$$\frac{\Delta e}{V} = \frac{m}{4} \frac{\Delta R}{nR} = \frac{m}{4} \frac{\Delta R}{R(1 + R_s/R)}$$

On the other hand, the change in the resistance of a strain gauge is related to strain ϵ and gauge factor G_F as

$$\frac{\Delta R}{R} = \epsilon G_F$$

Furthermore the strain ϵ is related to Young's modulus, E, and force F, through the relation

$$E = \frac{F/A}{\epsilon},$$

where A is the area of cross-section. Combining these equations, the bridge sensitivity $\dfrac{\Delta e}{F}$, may be written as

$$\frac{\Delta e}{F} = \left(\frac{mV}{4}\right)\left(\frac{1}{E(R + R_s)}\right)\left(\frac{G_F R}{A}\right).$$

Bridge sensitivity will remain independant of temperature provided the second and third bracketed terms are not influenced by it. The third term is independant of temperature if the gauges are so mounted that they are temperature compensated, and gauge factor is independant of temperature. The denominator of the second term taking into account temperature effect is expressed as:

$$E(R + R_s) = E(1 + c\,\Delta T)(R + R_s(1 + b\,\Delta T))$$
$$= E(R + R_s) + E\,\Delta T\,[bR_s + c(R + R_s)] + E\,cb\,R_s(\Delta T)^2$$

The influence of temperature on the second term $\dfrac{1}{E(R + R_s)}$ would be zero, for small temperature variations, when

$$b R_s + c (R_s + R) = 0$$

or $\quad \dfrac{R_s}{R} = -\dfrac{c}{b + c}.$

This indicates that temperature compensation may possibly be accomplished through proper balancing of the temperature coefficients of Young's modulus c, and electrical resistivity b. Because c is usually negative, and because resistance cannot be negative, it follows

$$b + c > 0.$$

Further the length of wire L required to achieve compensation is given by

$$L = -\frac{RA_1}{\rho}\left(\frac{c}{b + c}\right),$$

where ρ is the specific resistance and A_1 is the cross-section of the wire.

4.4 Torque measurement

The measurement of torque is necessitated in order to obtain load information necessary for stress or deflection analysis. The torque T may be computed by measuring the force F at a known radius r from the following relation:

$$T = Fr.$$

However, torque measurement is required for the determination of the mechanical power, either power required to operate a machine or power developed by the machine. The power is calculated from the relation:

$$P = 2\pi NT,$$

where N is the angular speed in revolutions per second. Torque measuring devices used in this connection are commonly known as dynamometers.

There are basically three types of dynamometers:

(i) *Absorption dynamometers:* They absorb the mechanical energy as torque is measured, and hence are particularly useful for measuring power or torque developed by power sources such as engines and motors.

(ii) *Driving dynamometers:* These dynamometers measure power or torque and as well provide energy to operate the devices to be tested. They are, therefore, useful for studying performance characteristics of devices such as pumps and compressors.

(iii) *Transmission dynamometers:* These are passive systems and are placed at an appropriate location within a machine or in between machines to measure torque at that particular location. These dynamometers are sometimes referred to as torque meters. Driving dynamometers are sometimes called as transmission dynamometers.

The first two types can be grouped as mechanical and electrical dyna-mometers and will be discussed as such here. However, the transmission type is treated separately.

MECHANICAL DYNAMOMETERS

These dynamometers are of absorption type. One of the most familiar and simple devices is the Prony brake as shown in Fig. 4.11. The

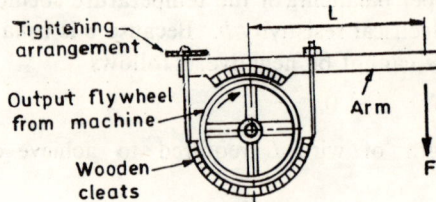

Fig. 4.11   Schematic of a Prony brake

mechanical energy is converted into heat through dry friction between the wooden brake shoes and the fly wheel of the machine. The torque exerted on the Prony brake is

$$T = FL,$$

where force F is measured by conventional force measuring instruments, say balances or load cells etc. The power dissipated in the brake is cal-culated from

$$P = \frac{2\pi NT}{60} = \frac{2\pi FLN}{60},$$

where force F (measured at arm L) is in Newtons, L is the length of reaction arm in meters, N is the angular speed in revolutions per minute, and P in watts.

The Prony brake is inexpensive, but it is difficult to adjust and maintain a specific load.

Various other types of brakes are employed for power measurements on mechanical equipment. The water brake, for example, dissipates the output energy through fluid friction.

ELECTRIC DYNAMOMETERS

Almost any kind of rotating electric machine can be used as a dynamo-meter but those specifically designed for the purpose are convenient to use. Electric dynamometers are of both absorption and driving type. The type of dynamometer to be used depends on the assigned task. For example if the machine to be tested is a power generator, the dynamo-meter must be capable of absorbing its power. On the other hand, if the machine to be tested is a power absorber, it must be capable of driving it. When the machine to be tested is a power transformer or transmitter, the dynamometer must provide both the power source and the load.

The electric dynamometers can be grouped into the following two classes:

1. d.c. dynamometers or generators, and
2. Eddy-current or inductor dynamometers.

In addition to these, ordinary electric motors and generators may also be used in dynamometry.

D.C. DYNAMOMETERS: The most versatile and accurate dynamometer is the cradled d.c. dynamometer. It is usable both as an absorption and as a transmission dynamometer. Hence it is most widely used for power and torque measurements on internal-combustion engines, small steam turbines, pumps etc A d.c. dynamometer is basically a d.c. motor with a provision to run it as a d.c. generator where the input mechanical energy on conversion to electrical energy can either be dissipated through a resistance grid or recovered for use. It is shown schematically in Fig. 4.12. When used as an absorption dynamometer, it performs as a d.c. generator. Cradling in trunnion bearing permits the determination of reaction torque. The torque is measured by measuring a balancing force (say by a load cell) at a fixed known moment arm extending from the body of the dynamometer. When used as a transmission dynamometer, it performs as a d.c. motor. It then measures the torque and power input to the machine for example a pump that absorbs power. Two good features of the d.c. dynamometer are its good performance at low speeds and ease of control. It can be adjusted to provide from zero to the so called base speed. Commercial dynamometers are equipped with controls for precise variation of the load and the speed of the machine.

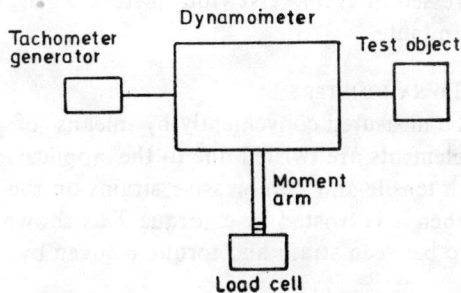

Fig. 4.12 A D.C. Dynamometer

EDDY-CURRENT DYNAMOMETER: An eddy-current dynamometer consists of one (or more) metal disc which is rotated in a magnetic field. The magnetic field is produced by passing current through coils which are attached to the dynamometer housing. The housing is mounted in trunnion bearing. The machine rotates the disc. As the disc rotates in the magnetic field, eddy currents are generated, and the reaction with the magnetic field tends to rotate the complete dynamometer housing in the trunnion bearings. The torque is measured with a moment arm extending from the dynamometer housing. A schematic of an eddy-current dynamometer is shown in Fig. 4.13. The power absorbed is carried away by water circulated through the air gap between the rotor and stator. The eddy-current dynamometers are only of absorption type and are used for the torque and power measurements of motors, small steam turbines etc. The eddy-current dynamometer can be easily controlled by varying current through the coils. It produces no torque at zero speed and only a small torque at low speeds. For a given capacity it is, however, of comparatively small size and possesses characteristics permitting good control at low speeds.

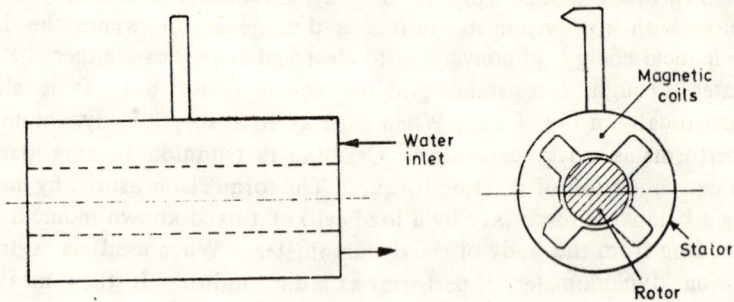

Fig. 4.13 An Eddy–current dynamometer

In practice torque and speed are measured and the power is computed from the relation

$$P = \frac{2\pi NT}{60}$$

where T is expressed in $N.m.$ (Newton meter), P in watts and N in revolutions per minute.

TRANSMISSION DYNAMOMETERS

Torque can be measured conveniently by means of solid or hollow tubes These elements are twisted due to the application of a torque. There exist both tensile and compressive strains on the surface at 45° to the tube axis when it is twisted by a torque T as shown in Fig. 4.14(a). The relationship between strain and torque is given by

$$\epsilon_{45°} = \pm \frac{Tr_0}{\pi G(r_0^4 - r_1^4)},$$

where G is the shear modulus of elasticity, r_0 and r_1 are the outside and inside radii of the tube. The angle of twist ϕ is given by

$$\phi = \frac{2Tl}{G(r_0^4 - r_1^4)},$$

where l is the length of the tube.

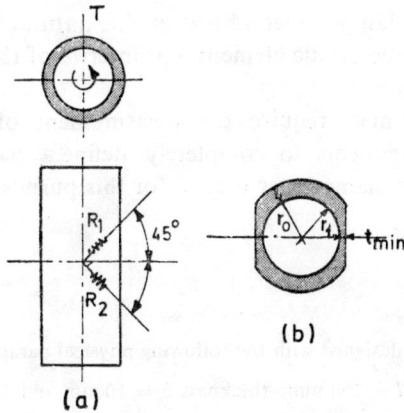

Fig. 4.14 Torque tubes

The sensitivity to strain can be increased by thinning the tube wall at the gauging sections as shown in Fig. 4.14(b). The strain will be increased by the ratio of the initial thickness $(r_0 - r_1)$ to the thickness t_{min} at the gauging section. The strain is measured by fixing strain gauges to the tube. The gauges are so mounted as to be insensitive to any axial or bending load and also to any temperature changes. They respond only to the twist.

In some cases, other forms of elastic elements such as bars of rectangular cross-section etc. are used. The strain gauges are mounted on torque sensing element as shown in Fig. 4.15. The strain gauges are so mounted as to respond to twist only. Sometimes optical methods are used as a coupling between driving and driven machines, or between any two portions of a machine. In cases where strain gauges are used as secondary transducers, electrical connections are made through slip rings, with a provision to lift the brushes when they are not in use, thus minimising wear. However, slip rings are subject to wear and may present maintenance problems when permanent installations are required. Attempts are therefore made to develop electrical torque meters which do not require electrical connections to the moving shaft thereby completely eliminating slip rings. Inductive and capacitive transducers are used to accomplish this.

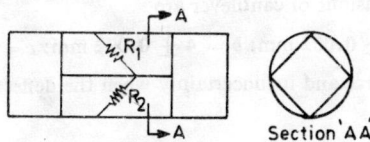

Fig. 4.15 Square shaft torque element

The dynamic response of elastic elements used for torque measurement is essentially damped second order. The natural frequency depends on the stiffness of the elastic element and intertia of the parts connected at either end.

Some situations may require the measurement of three force and three moment components to completely define a particular condition. Multi component dynamometer is used for this purpose.

Exercises

1. A proving ring is designed with the following physical parameters:

   Overall diameter $d = 160$ mm: thickness $b = 10$ mm and depth $h = 25$ mm.

   It is constructed of steel ($E = 2 \times 10^{11}$ N/m²). A micrometer with an uncertainty/ of ± 2 μm is used for deflection measurement.
   (a) Assuming that the dimensions of the ring are exact, calculate the applied load when the uncertainty in load is 2%.
   (b) If the dimensions are:

   $$d = 160 \pm 0.2 \text{ mm}: b = 10 \pm 0.01 \text{ mm}: \text{and } h = 25 \pm 0.01 \text{ mm}.$$

   Calculate the percentage uncertainty in the load when the deflection is 0.2 mm.

2. A torque is applied to a t··que tube made of steel ($G = 8 \times 10^{10}$ N/m²) and of physical dimensions as:

   $r_0 = 16 \pm 0.01$ mm, $r_1 = 12 \pm 0.01$ mm and the length $L = 125 \pm 0.03$ mm.

   The angle of twist ϕ is $1.50°$ and its uncertainty is $\pm 0.05°$.
   (a) Calculate the nominal value of the impressed torque and its uncertainty.
   (b) Calculate the 45° strains.

3. Answer the following:
   (a) Is a torque tube with four strain gauges mounted at 45° to the axis insensitive to the axial load?
   (b) Does a hollow torque tube have a higher natural frequency than a solid one for a given strain or torque sensitivity?
   (c) Does a hollow cylindrical cantilever bar have a higher natural frequency than solid one for a given strain or load sensitivity ?
   (d) Is a square shaft stiffer in bending than a round one for an equivalent strain or torque sensitivity?

4. Design a load cell having an electrical output, and a resolution of 0.1 gm when used with a read-out device sensitive to 10^{-5} volt. Estimate the natural frequency of the cell.

5. A cantilever of spring steel ($E = 2 \times 10^{11}$ N/m²) is used for the force measurement. The deflection is measured with a micrometer having a sensitivity of ± 1 μm. The dimensions of cantilever are

   $$L = 25 \pm 0.002 \text{ mm}: b = 4 \pm 0.002 \text{ mm}: c = 0.75 \pm 0.001 \text{ mm}:$$

   Calculate the force and its uncertainty when the deflection is 2 mm.

5

Measurement of Strain and Stress

5.1 Introduction

When a stress is applied to a body, it gets deformed and these deformations are related to the applied stress. Under the state of loading it is very desirable to find out the actual stress distribution in the body. The evaluation of stress distribution in the body is known as stress analysis and includes the determination of kind, magnitude and direction of the stress. Theoretical stress analysis is possible for bodies of simple geometries; however, for complicated geometries many assumptions are made and the mathematics is too complicated. For this reason one resorts to experimental stress analysis. For stress analysis, the stress is not directly measured but obtained indirectly from the measurement of deformations or strains using strain-stress relations. Therefore an experimental stress analysis is not 100% experimental but involves the use of theoretical relations.

There are a number of methods available for measuring strain. The following methods are discussed in this chapter:
  (i) Mechanical methods
  (ii) Opto-mechanical methods
  (iii) Electrical strain gauges
      (a) capacitance gauges
      (b) inductive gauges
      (c) piezo-electric gauges
      (d) resistance gauges
  (iv) Grid method
  (v) Moiré Fringe technique
  (vi) Interferometry
  (vii) Photo-elasticity
Holographic methods currently in vogue for specialised applications are discussed in a separate chapter.

5.2 Mechanical methods

Since the magnitude of deformation is very small, some kind of magnification is to be provided for the measurement. Prior to

1930, all instruments used for the measurement of deformation were of mechanical nature and amplification was achieved with mechanical elements like wedges, screws, levers, gears and their combinations. The length of the gauge was about 250 mm and hence only a value of deformation averaged over such large lengths was experimentally obtained. Therefore, these mechanical gauges were not suitable for the measurement of steep gradients of strain. Further it is rather difficult to design a mechanical system which gives a precise and desired magnification. Various other factors like friction, lost motion, the weight and inertia, and the flexibility of parts hinder accurate measurements by these instruments. These instruments cannot be used for the study of dynamic strains as are obtained in impact loading, etc.

Figure 5.1 shows one such mechanical gauge using lever and dial gauge combination for amplification. Another mechanical gauge where the amplification is achieved in an ingenious way is presently available. This is known as 'microkator' and makes use of a double helix, oppositely twisted with a pointer attached in the centre. The gauge is very sensitive; it can sense displacements in the range of $0.2\ \mu$m. It has a fair dynamic response.

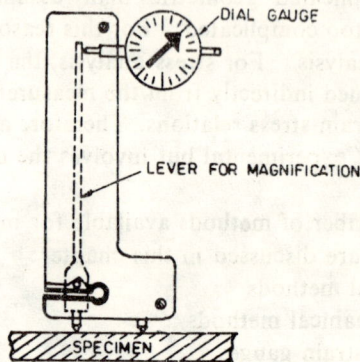

Fig. 5.1   Mechanical strain gauge

5.3   Opto-mechanical methods

Optical levers are used for magnification in the opto-mechanical gauges. Because of high magnification achievable, these gauges are of smaller length typically of 50 mm. One such gauge is the Tuckerman gauge shown in Fig. 5.2. It consists of two separate sub-systems (i) an extensometer and (ii) an auto-collimator. The extensometer comprises of a right angle roof prism and a tungsten-carbide rocker called as the Lozenge by the manufacturer. The auto-collimator provides a beam of collimated light and receives it back after reflection from the prism-rocker assembly of the extensometer. The extensometer is mounted on the specimen and the right angle roof prism is adjusted such that the beam reflected from prism-rocker assembly makes an image of the graticule at a reference point as viewed through an eye-piece. When the specimen is deformed, say elongated along the length of the gauge, the rocker edge will move outwardly tilting its mirror surface by an angle ϕ. The reflected beam from prism rocker assembly will be deflected by an angle of 2ϕ, which results in the shift of the image by $2f\phi$, where f is the focal length of the auto-collimator. Therefore, the linear shift at the focal plane of the auto-collimator is dependent on the tilt of the rocker and the focal length of the auto-collimator. From the design data, the shift of the image is calibrated as the strain. An important feature of this sort of instrument is that there need not be a fixed relationship between the positions of the auto-collimator and the extensometer.

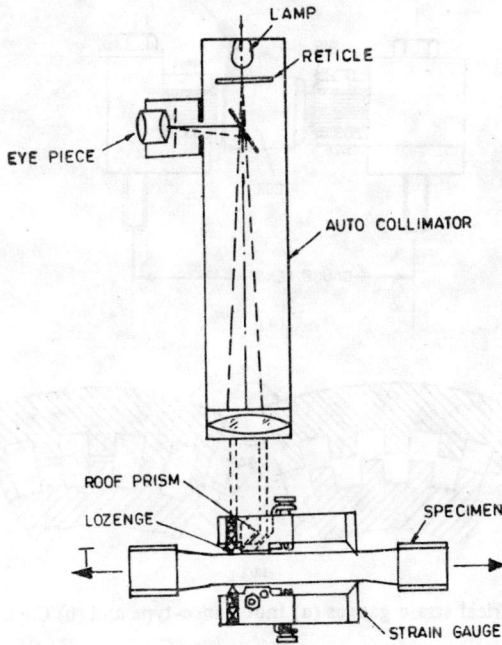

Fig. 5.2 Schematic of mechanical strain gauge with optical amplification

5.4 Electrical methods

The methods which measure the change of some electrical quantities arising due to deformation (strain) in the body fall under electrical methods. These include (i) capacitive gauges, (ii) inductive gauges, (iii) piezo-electric gauges, and (iv) resistance gauges. The electrical methods of measuring strain possess the advantage of high sensitivity and ability to respond to dynamic strains. Both the capacitive and inductive type gauges are of large mass and size, and are used only for some special applications. These gauges are sometimes used as load indicators, mounted directly on the machine frame. In one application an inductive gauge is mounted on the frame of a rolling-mill for the measurement of rolling loads in steel mill. These gauges are quite rugged and maintain the calibration over long period of time.   Figure 5.3 (a) illustrates all inductive gauge for general purpose application. A deformation results in the variation of air gap and hence the inductance.   A linear differential transformer can also be used as an inductive gauge. Figure 5.3(b) shows a capacitive type strain gauge which is used in a torque meter. Torque carried by an elastic member causes a shift in the relative positions of the teeth, thereby changing the effective area and hence the capacitance. The changes in inductance or capacitance due to strains caused by loading are calibrated in terms of strain.

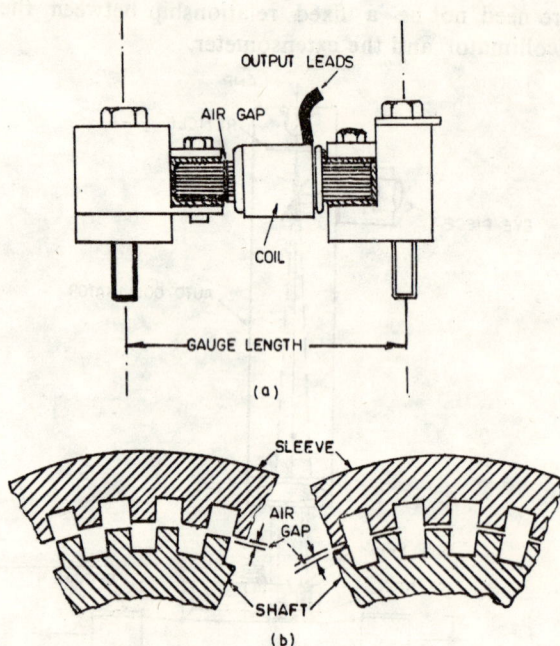

OUTPUT LEADS

AIR GAP

COIL

GAUGE LENGTH

(a)

SLEEVE

AIR GAP

SHAFT

(b)

Fig. 5.3   Electrical strain gauges (a) Inductance–type and (b) Capacitance type

Piezo-electric strain gauges are mainly used for studying dynamic inputs.   Some transducers having very high internal resistance and proper circuitry may be used to measure slow varying inputs,   but rarely steady inputs.   The gauge is cemented to the specimen.   The voltage output developed when the   gauge is deformed along with the specimen is taken as a measure of strain.   The piezo-electric gauges are equally sensitive to strains in lateral directions and have a very high output sensitivity.   One of the piezo-electric materials that is used for this purpose is barium-titanate.   Wafers of barium-titanate of 0.25 mm thickness with suitable electrodes are bonded to the specimen with Duco cement.

Due to their good dynamic response, stability, range of available size, ease of data presentation and processing etc., resistance strain gauges are widely used for stress analysis.   The resistance   strain gauges work on the principle of piezo-resistivity: the resistance of a wire conductor changes when it is strained. The change in the resistance bears a definite relation with the strain or the applied stress. The resistance strain gauges are of two types:

  (i) unbonded strain gauges, and
  (ii) bonded strain gauges.

The unbonded strain gauges are made of a high-tensile resistance wire (commercially referred to as alloy 479, containing 92% Pt and 8% W) of about 0.025 mm diameter and of about 25 mm in length. Two to twelve loops of the wire are attached to both a stationary frame and a movable platform with the help of pins made of electrically insulating material. Relative motion between the stationary frame and the platform is possible as guided by flexure plates. One such construction is shown in Fig. 5.4. The resistance wire is preloaded so that it could be used to measure compressive strain. The four resistance wires in this construction are so wired that the bridge acts as a full bridge as shown in Fig. 5.4. Unbonded strain gauges are mainly used as elements of force and pressure transducers, and accelerometers rather than for strain measurement. On the other hand, bonded type of

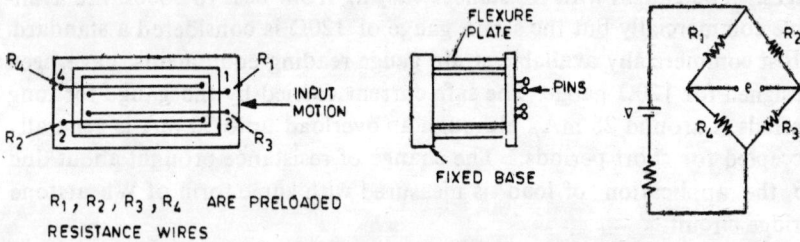

Fig. 5.4 Unbonded strain gauge

strain gauges are cemented to the specimen. When properly cemented, they effectively form part of the surface and undergo the same strain. A bonded type of strain gauge consists of either a length of a fine metal wire of approximately 0.025 mm diameter, or a metal foil of 0.005 mm thick or a whisker of a semiconducting material as a resistive element. To reduce the length of the gauge while retaining its sensitivity ($\propto R$), the wire, or the foil is usually formed in a grid pattern. Due to the high sensitivity of the semiconductor strain gauge, a single whisker is usually used. These are shown in Fig. 5.5. The resistance element is formed on a suitable base, for example.

  (i) Paper base: the wire is wound around a paper or sandwiched between the paper. Leakage resistance is of the order of 100MΩ.
  (ii) Bakelite base: the wire and foil gauges used for high temperature applications are formed on bakelite base.
  (iii) Plastic and rubberised stripable base: mostly foil gauges are formed on plastic or rubberised base where from they are stripped for cementing to the test member.

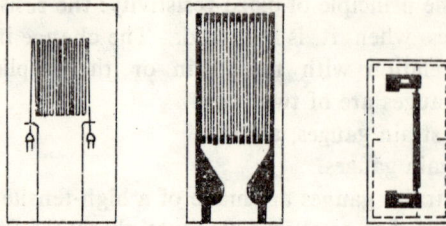

Fig. 5.5 Types of electrical resistance strain gauges

The paper base strain gauge may be cemented with Duco cement to the test member under a pressure of 0.15 kg/cm². It cures in about 7–8 hours. Phenol resin is employed for bakelite base strain gauges which are held to the specimen with a pressure of 1.8 kg/cm² during curing. Curing lasts 24 hours. Epoxy resin may be used for plastic base strain gauges.

The most common metals for the manufacture of strain gauges are (i) an alloy of copper and nickel (55% Cu and 45% Ni), and (ii) an alloy of nickel, chromium and iron with other elements in very small percentages. The gauges with resistances varying from 60Ω to 5000Ω are available commercially but the strain gauge of 120Ω is considered a standard. Most commercially available strain gauge reading equipments have been designed for 120Ω gauge. The safe current carried by the gauge for long periods is around 25 mA, although an overload upto 50 mA is generally accepted for short periods. The change of resistance brought about due to the application of load is measured with some form of Wheatstone bridge circuit.

Gauge size and shape are very important. A narrow long gauge, 10 to 25 mm long, can be employed to measure strain on the test member if the stress distribution is fairly uniform. In regions of likely stress concentration and steep strain gradients, a gauge of much shorter length must be used. Gauges of even 0.4 mm effective lengths are available.

A gauge will only react to strain parallel to the length of the wire. If the strain is not parallel to the gauge length, then only the component of strain that lies in the direction of gauge length will be measured. Therefore,

(i) single element gauges are used in uniaxial fields or for making rosettes. If the gauge is used as prescribed by the calibration procedure, strain measurement will be accurate otherwise a correction for cross-sensitivity is to be applied.

(ii) two element gauges are used in a biaxial stress field, in which the directions of principal stresses are known and only their magnitudes are to be determined. In the region of high stress concentration, the single element gauges (grids) are stacked one on top of the other but insulated from each other so that the gauge is effectively of a smaller size.

(iii) three element rosette gauges are used for the study of a general biaxial stress field, in which neither the direction nor the magnitude of the principal stresses are known. The choice of stacking or single plane style is determined by the nature of strain gradient at the point of consideration where the gauge is to be mounted. Figure 5.6 illustrates a single element, two elements stacked over each other, three elements 60° rosette in plane style and 90°–45° rosette in overlapping style configurations of strain gauges.

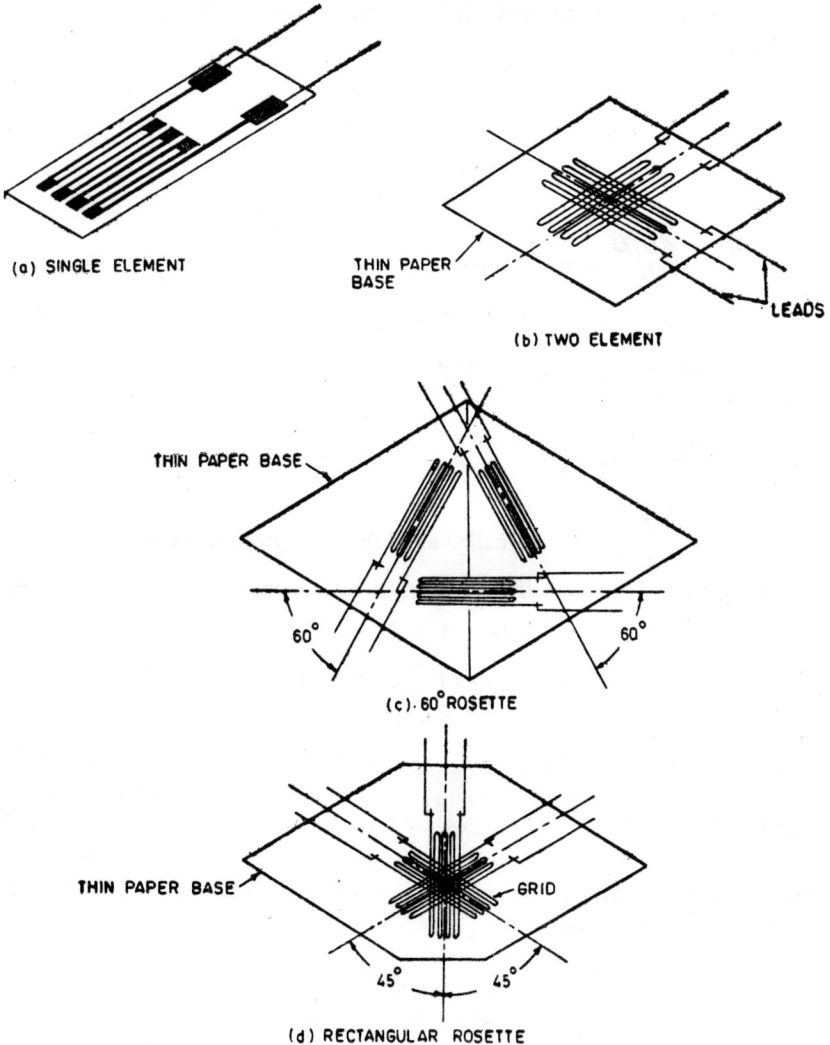

Fig. 5.6 Types of resistance gauge configurations

ANALYTICAL THEORY OF STRAIN GAUGE: PIEZO-RESISTIVE EFFECT

A simplified approach to the theory of strain gauge is presented here.

The resistance R of a wire of length L and area of cross-section A is given by

$$R = \frac{\rho L}{A},$$

where ρ is the specific resistivity. On differentiation one obtains

$$\frac{dR}{R} = \frac{d\rho}{\rho} + \frac{dL}{L} - \frac{2dD}{D},$$

where $A = CD^2$, C is a constant, its value being one for square cross-section of dimension D, and $\pi/4$ for circular cross-section of diameter D.

Defining

$$\frac{dL}{L} = \epsilon_a = \text{axial strain,}$$

$$\frac{dD}{D} = \epsilon_l = \text{lateral strain, and}$$

$$\frac{-dD/D}{dL/L} = \mu = \text{Poisson's ratio,}$$

the above equation can be recast as

$$\frac{dR/R}{dL/L} = 1 + 2\mu + \frac{d\rho/\rho}{dL/L} = G_F.$$

or $$\frac{dR}{R} = G_F \, \epsilon_a.$$

The term $\dfrac{dR/R}{dL/L}$ ($= G_F$) is called axial sensitivity of gauge factor or gauge factor itself. If the specific resistivity does not depend on the strain then

$$G_F = 1 + 2\mu.$$

Usually the value of Poisson's ratio for most metals is 0.3. Therefore,

$$G_F = 1.6$$

In fact the measured values of gauge factor for metals vary from -12 for Ni to 0.47 for manganin (an alloy of Ni, Cu and Mn). This is perhaps due to the fact that the behaviour of ρ with strain for very thin wires used for strain gauges is not so well understood. Table 5.1 gives the gauge factor, temperature coefficient of resistance etc. for some materials used for strain gauge work.

Table 5.1 shows a very wide variation of the gauge factor. Therefore the gauge factor for each composition type is either to be measured before using the gauge or is supplied by the manufactures which usually is the case.

If the resistance gauge is strained to the extent that its element is operating in the plastic region, then $\mu = 0.5$ and hence $G_F = 2.0$. For most commercial strain gauges the gauge factor is the same for both compressive and tensile strains.

TABLE 5.1

Gauge meterial	Composition %		Gauge factor G_F	Temperature coefficient of resistance $(\mu\Omega/\Omega°C)$	Remarks
Advance	57 43	Cu Ni	2	10.8	G_F is constant for wide range of strain used below 250°C.
Nichrome	80 20	Ni Cr	2	396	High temperature co-efficient of resistance
Isoelastic	36 8 0.5 55.5	Ni Cr Mo Fe	3.5	468	Mostly used for dynamic strain measurements
Platinum alloys	95 5	Pt Ir	5.1	1260	Used for high temperature above 550°C
Ni	100	Ni	—12	43,200	
Semiconductors	—		—140 to 175	90,000	Not suitable for large strain measurement-limited range

Cross sensitivity

As has been pointed out earlier, a strain gauge is formed in a grid pattern in order to shorten its length while retaining its sensitivity. Thus although most of the gauge element wire is parallel to the length (axis) of the gauge, a certain length of the wire is unavoidably placed transverse to the gauge axis. This length of wire will react to the strain perpendicular to the gauge-axis and gives rise to cross-sensitivity G_{FC}. The ratio G_{FC}/G_F is generally about 2%. In foil type gauges, the effect of this length is considerably minimised by decreasing the resistance of the cross length, the foil at the ends is usually of larger surface area and hence of low resistance.

The manufacturer's calibration is carried out in a uniaxial stress field, so that if the gauge is used in service in the same manner, cross-sensitivity may be ignored. However when the strain gauge is used in a biaxial stress field, the readings will be influenced by the transverse strain and a correction should, therefore, be applied to obtain the true strain along the gauge axis. When a pair of strain gauges are used in a biaxial field with their axes along the principal stress directions, the true strains ϵ_1 and ϵ_2 along the two perpendicular directions are given by

$$\epsilon_1 = \frac{1 - \mu\eta}{1 - \eta^2}(\epsilon_1' - \eta\epsilon_2'),$$

and $\qquad \epsilon_2 = \dfrac{1 - \mu\eta}{1 - \eta^2}\,(\epsilon_2{}' - \eta\epsilon_1{}')$,

where $\epsilon_1{}'$ and $\epsilon_2{}'$ are the measured strains along these directions and $\eta\ (= G_{FC}/G_F)$ is the ratio of transverse sensitivity to the axial sensitivity of the gauge.

Effect of Temperature and Humidity

After the strain gauge has been mounted on a test member, variation of room temperature influences the gauge readings in the following three ways:

(i)   the gauge factor of the strain gauge is effected by temperature owing to creep, i.e. the gauge factor varies with time for a constant applied stress. Constantan wire gauge with bakelite base and phenol resin cement has minimum creep up to temperatures of 200°C,

(ii) the resistance of the strain gauge element will vary with a change in the temperature, and

(iii) an apparent strain will be induced in the gauge due to the differential expansion between the test member and the gauge bond material.

The influence of temperature on the gauge readings can be mathematically expressed by the following equations:

$$\frac{\Delta R}{R_0} = \epsilon G_{F_0}\,(1 + mG_{F0}\Delta\theta) + \beta\Delta\theta + (a_2 - a_1)\Delta\theta$$

where G_{F_0} is the gauge factor at room temperature θ_0, mG_{F_0} the temperature coefficient of gauge factor, β the temperature coefficient of resistance of gauge wire, a_1 and a_2 are the coefficients of linear expansion of test member and the gauge wire respectively, and $\Delta\theta = \theta - \theta_0$; θ is the temperature of the gauge. If the temperature coefficient of resistance of the test member is more than that of the gauge wire $(a_1 > a_2)$ at a given temperature, the strain gauge suffers tensile strain. On the other hand if $a_2 > a_1$, it will experience compressive strain.

Partial or complete temperature compensation over a limited range of measured strain values is achieved by:

(i)   making

$$\beta + (a_1 - a_2) = 0.$$

For metals β is positive. Therefore this equation can be satisfied by suitably choosing the gauge material for a definite test member. It is however assumed that the gauge factor is independent of temperature.

(ii) using a series combination of two 'opposing' wire materials such as constantan and nickel (dual gauges), the latter having a negative gauge factor. This method provides a complete compensation at least over a limited range of strain values.

(iii) locating a temperature-sensitive resistor close to the gauge and connecting it electrically in series with the bridge supply line so that the sensitivity (output voltage/input strain) remains unaltered. This method has been discussed in Chapter 4.

The most common and practical method of compensating temperature errors is either by the use of dummy gauges or by employing wire resistance gauges in push-pull pairs. Examples of the various arrangements for compensating temperature errors are given in a later section of this chapter.

Humidity is another factor which may seriously impair the performance of strain gauges. Humidity causes corrosion of gauge wire resulting in an increase in its resistance. The effect of humidity on the performance of the strain gauge can be minimised by giving a coat of wax on the bonded strain gauge.

STRAIN-GAUGE SENSITIVITY EQUATION

Strain induced resistance change of the strain gauge is measured with some form of a Wheatstone bridge. Figure 5.7 shows a Wheatstone bridge circuit with four active gauges of resistances R_1, R_2, R_3 and R_4, and excited by a voltage V. Initially the bridge is balanced, i.e.

$$\frac{R_1}{R_4} = \frac{R_2}{R_3}$$

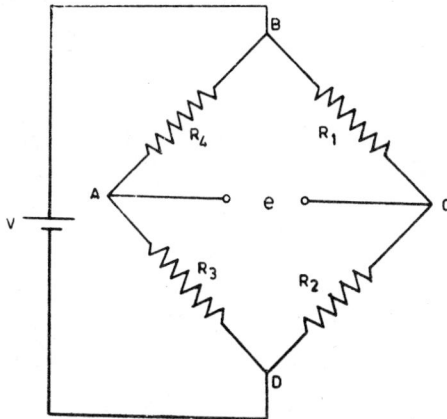

Fig. 5.7 Wheatstone bridge circuit

In practice it is very difficult to achieve null balance by proper selection of the resistances. Usually series or parallel balancing network is used. These might modify the sensitivity of the bridge.

On application of stress, the active gauges will undergo a change of resistance dR_i ($i = 1, \ldots, 4$) and hence a potential difference e will be developed across AC. The potential difference e is given by

$$e = V\left[\frac{R_2\,dR_1}{(R_1 + R_2)^2} - \frac{R_1\,dR_2}{(R_1 + R_2)^2} + \frac{R_4\,dR_3}{(R_3 + R_4)^2} - \frac{R_3\,dR_4}{(R_3 + R_4)^2}\right],$$

If it is assumed that all the four strain gauges have same nominal resistance i.e. $R_1 = R_2 = R_3 = R_4 = R$, the output voltage e can be expressed in terms of the gauge factor and the strains as

$$e = \frac{VG_F}{4}(\epsilon_1 - \epsilon_2 + \epsilon_3 - \epsilon_4) = \frac{VG_F}{4}\epsilon_{net},$$

where ϵ_{net} is the net strain. This is known as the bridge sensitivity equation. The sensitivity S_b of the bridge is defined as

$$S_b = \frac{e}{\epsilon} = \frac{VG_F}{4}\left(\frac{\epsilon_{net}}{\epsilon}\right),$$

where ϵ is the strain with one active gauge.
For a constant voltage bridge it is expressed as

$$S_b = \frac{i_g RG_F}{2}\frac{\epsilon_{net}}{\epsilon}.$$

This would be $2i_g RG_F$ for the full bridge. i_g is the current through the gauge of resistance R. Based on the equation for e and requirements of temperature compensation, some arrangements for mounting strain gauges for some specific purposes are now described.

MOUNTING OF GAUGES

1. *Positioning of gauges to measure axial strain only*

Figure 5.8 (a) and (b) show strain gauges mounted on a bar and the schematic of the bridge circuitry respectively. The gauges R_1 and R_3 are

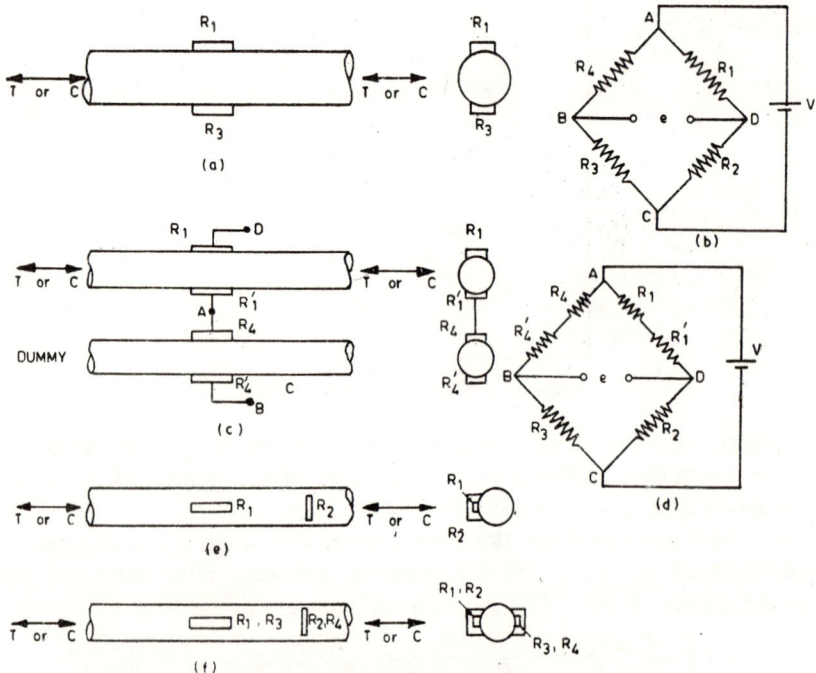

Fig. 5.8 Positioning of strain gauges to measure axial strain

mounted axially on the bar which is subjected to either tensile or compressive stress, gauges R_2 and R_4 are located where they remain unstrained. Therefore,

$$\epsilon_2 = \epsilon_4 = 0,$$

and
$$\epsilon_1 = \epsilon_3 = \epsilon.$$

Therefore, the voltage e developed across DB is $VG_F\epsilon/2$. The arrangement provides double the sensitivity compared to that achievable with a single active gauge. Further the arrangement is insensitive to bending stress, as one of the gauges is compressed ($-\Delta R$) while the other is elongated thus cancelling the effect of bending stress provided they are bonded diametrically opposite to each other.

If temperature compensation along with the removal of the influence of bending strain is required, the gauges are mounted as shown in Figs. 5.8 (c) and (d). The resistances R_4, R_4' are identical to R_1, R_1' and are used as dummy gauges for temperature compensation. The resistances R_1 and R_1' are bonded diametrically opposite to each other for annulling the effects of bending strain.

Another very simple arrangement but mostly recommended for foil gauges ($\simeq$ zero cross sensitivity) is shown in Figs. 5.8 (e) and (f). In this arrangement, temperature compensating gauges are mounted perpendicular to the axis of the bar and hence are not subjected to the axial strains. Bending strains are avoided by mounting R_1, R_3 and R_2, R_4 diametrically opposite to each other. The sensitivity is double and the bridge is said to be working as half bridge.

2. *Positioning of gauges to measure bending strains only*

Figures 5.9 (a) and (b) show two arrangements to measure bending strains only. The gauges R_1 and R_2 are mounted diametrically opposite to each other; the arrangement is inherently insensitive to temperature

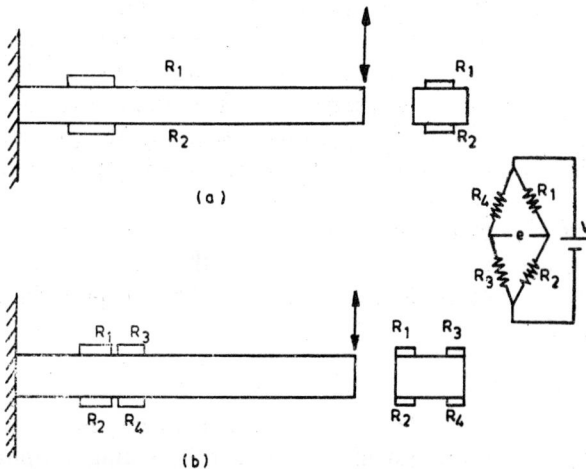

Fig. 5.9 Mounting of gauges to study bending

effects.   In the arrangement of Fig. 5.9 (a) the bridge works as a half bridge.   The bridge can be made to work as full bridge if the gauges R_1 and R_3 are mounted on one side and the gauges R_2, R_4 on the diametrically opposite side of the test member as shown in Fig. 5.9 (b).   In full bridge, the sensitivity is four times that achievable with a single active gauge.

3.   *Positioning of gauges to measure torsional strains*

A cylindrical bar subjected to torsion has principal strain directions at 45° to the longitudinal axis of the bar.   Both the bending and axial strains can be eliminated by mounting the strain gauges in an arrangement shown in Fig. 5.10. Since the principal axes of strain due to torsion are at 45° to the longitudinal axis of the cylinder and the strain gauges R_1 and R_2 are mounted as shown in Fig. 5.10, the resistance of R_2 will increase ($+ \Delta R_2$) and that of R_1 will decrease ($- \Delta R_1$), thus pulling the effect.   However, resistance changes due to axial strain and temperature will be positive for both the gauges R_1 and R_2, and thus compensate each other.   Similar arguments hold good for gauges R_3 and R_4.   Further R_1, R_2 and R_3, R_4 are diametrically placed, thus the effect of bending will be compensated.

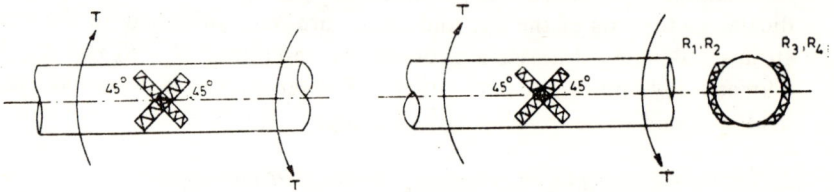

Fig. 5.10   Positioning of gauges to measure torsion

4.   *M ...unting of gauges in pressure pickups*

A favourable situation occurs in a diaphragm type pressure pickup. The diaphragm is rigidly clamped at the periphery and is uniformly loaded.   Both compressive and tensile strains coexist.   The gauges are mounted such that the bridge acts as a full bridge and temperature effects are compensated by push-pull action of the gauges.   Details of the diaphragm type pressure pickups are given in the next chapter.

CALIBRATION OF STRAIN GAUGES

Strain gauges may be used under both the static and dynamic conditions.   Hence both static and dynamic calibration procedures shall now be discussed.

Static calibration

Static calibration refers to a situation where an accurately known input is applied to the system and the corresponding output measured, while all other inputs are kept constant at some value.   This procedure often cannot be applied in the bonded strain gauge work because of the nature of the gauge (transducer). Normally, the gauge is bonded to the test member to measure the strains. Once the gauge is bonded, it can be hardly transferred to known strain situation for calibration. Therefore when the gauge is used to experimentally measure the strain, some other approach to the calibration problem is required. The method of calibration is based on the fact that the value of both the gauge factor and gauge resistance are known accurately.

Resistance strain gauges are manufactured under carefully controlled conditions. The gauge factor for each lot of gauges is provided by the manufacturer with an indicated tolerance of about $\pm 0.2\%$. Since both the gauge resistance and gauge factor are known, a simple method of calibration is to determine the response of the system to the introduction of a known small resistance change at the gauge and calculate the

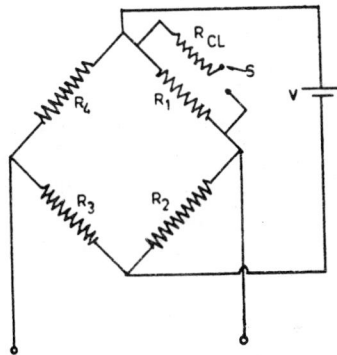

Fig. 5.11 Calibration method

equivalent strain therefrom. A number of precision high resistances are provided parallel to the gauge and a small change in the resistance of the gauge is obtained by shunting one of the precision resistors. Figure 5.11 shows an arrangement that may be used for calibration. When the switch S is closed, the resistance of the arm containing R_1 is changed by a small amount ΔR. The resistance change ΔR is given by

$$\Delta R = \frac{R_1^2}{R_1 + R_{CL}},$$

where R_{CL} is the calibration resistance. The equivalent strain is thus

$$\epsilon_{CL} = \frac{1}{G_F} \cdot \frac{\Delta R}{R_1} = \frac{1}{G_F} \cdot \frac{R_1}{R_1 + R_{CL}}$$

A number of equivalent strain values can be obtained by shunting different calibration resistors across R_1. A graph between resistance change ΔR and the equivalent strain then can be plotted. The strain corresponding to the measured resistance changes when the gauge is bonded on the test member is read off from the ϵ_{CL} vs ΔR plot.

Dynamic calibration

Dynamic calibration of the strain gauge is carried by fastly shunting the gauge R_1 repeatedly. This is achieved by replacing the manual calibration switch S with an electrically driven switch. This electrically driven switch is referred to as 'chopper', which makes and breaks the contact 60 to 100 times per second. When the output of the bridge is displayed on CRO screen or recorded, the trace obtained is found to be a square wave. The step in the trace represents the equivalent strain calculated from:

$$\epsilon_{CL} = \frac{1}{G_F} \cdot \frac{R_1}{R_1 + R_{CL}}$$

The dynamic response of the bonded strain gauge i.e. of faithfully reproducing resistance changes corresponding to strain variations of the test member upto 50 KHz is extremely good. The dynamic behaviour of the strain gauge is that of a zero-order system.

MEASUREMENT OF STRAINS

The bonded strain gauges are used to measure extremely small displacements (or strains). However they can be used to measure large displacements as well provided an intermediate elastic element is used. For example, gauges may be bonded to a cantilever and an unknown displacement is applied at its end. The method does not require calibration of the strain gauge as outlined above, rather the overall system can be calibrated by giving known displacements to the end of the cantilever and studying the system's response.

Strain induced resistance changes are measured by a Wheatstone bridge circuitry. The bridge can be excited either by a d.c. or an a.c. source. Further, the strain to be measured could be of static, slowly varying or of dynamic nature. For static strains, a galvanometer can be used for measurements. However, its use for dynamic strains is restricted due to the high inertia of its moving parts. Therefore a pen recorder or a CRO which possesses the required dynamic range is used for dynamic strain measurements. Further, the output of the bridge is very small, and hence a high gain d.c. amplifier for static or slowly varying strains is required. A schematic of a d.c. excited measurement system is shown

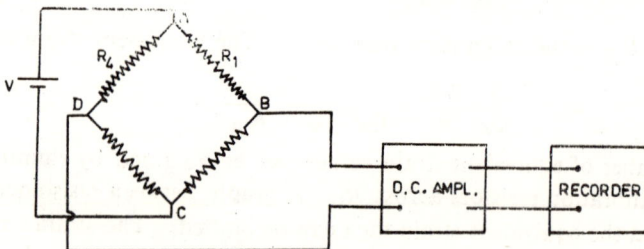

Fig. 5.12   Recording of slow varying static strains

in Fig. 5.12. Fabrication of a high gain d.c. amplifier to amplify such low level signals of very low frequency poses the following problems:

(i) Contact potentials and thermo emf's in the various parts of the circuit may be comparable to the unbalance voltage due to strain and are also amplified along with the signal.

(ii) Small changes in the terminal potentials at different stages due to d c. drift will be of the same order of magnitude as of the unbalance voltage and will be amplified along with the signal.

(iii) Stray potential of 50 Hz from the vicinity of the long cables used for measurements may appear as noise and influence the measurements.

Essentially there is no inbuilt discrimination in a d.c. amplifier which may discriminate between the desired signal and undesired voltages that are developed at various stages of the circuit. On the other hand an a.c. amplifier though suitable for dynamic measurements cannot be used for static or slowly varying signals except when chopper is used in the circuit. A method using a carrier frequency for both the slow varying and dynamic strains is described next.

a.c. excited bridge

Figure 5.13 shows an a.c. excited bridge. The a.c. source is connected across the terminals A and C of the bridge, and the output across the terminals B and D is fed to an a.c. amplifier. The frequency of the a.c. supply, often called a carrier, is usually between 50 Hz to 10 KHz, although frequencies upto 50 KHz have been used. The working of an a.c. excited bridge is explained with reference to a d.c. excited bridge

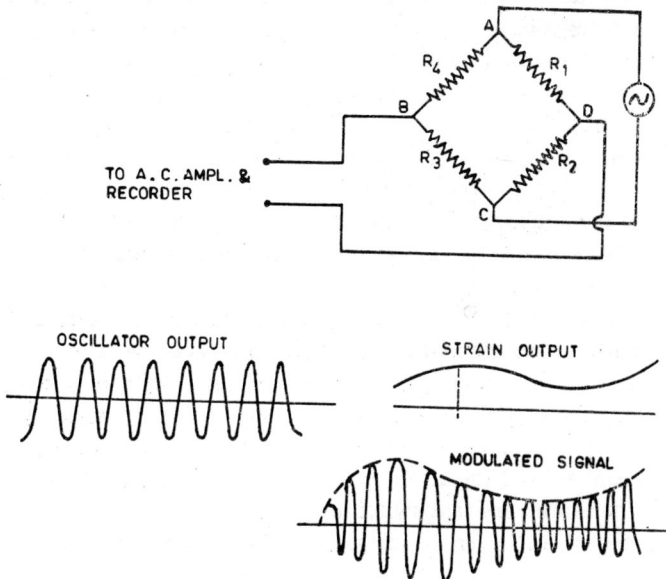

Fig. 5.13 a.c. excitation of bridge

already discussed. When the bridge is balanced, the potential difference across BD is zero. If the resistance R_1 is increased by ΔR_1 due to the application of load, the potential of terminal B will be higher than that of D. Conversely if the resistance R_1 is decreased, the reverse will hold good. Therefore tension and compression strains produce potentials of opposite polarity at the bridge output. However for an a.c. excited bridge the voltage at the terminals A and C is alternately positive and negative, and hence any unbalance of the bridge would result in an a.c. output. Indeed both tensile and compressive strains will provide an a.c. output. It can, however be shown that the tensile and compressive strains produce outputs of opposite phase; the tensile strain producing an output in phase with the supply.

The effect of unbalancing the bridge by subjecting the gauges to strains is to amplitude modulate an a.c. signal of carrier frequency. Therefore, the output of the bridge can be amplified by an a.c. amplifier irrespective of the nature of the strain i.e. static or dynamic. For the measurement of static strains, the frequency of carrier need not be high. It should, however, be noted that the bridge output voltages are unsuitable for recording and analysing into strain value due to the following reasons:

(i)   carrier frequency operation requires processing system of higher frequency response than would be necessary for the strain signal alone.

(ii) presence of carrier frequency tends to confuse interpretation of dynamic strains, and

(iii) both tension and compression strains cause an increase in the amplitude of carrier voltage at the bridge output terminals.

These limitations of an a.c. excited bridge for strain measurements can be eliminated in either of the two ways:

(i)   The first method consists of rectifying and filtering the amplitude modulated carrier signal after amplification, thus eliminating the carrier frequency. In order to distinguish between the tensile and compressive strains, the bridge is intentionally unbalanced initially to the extent that the maximum anticipated strain will not bring it back to the balance point. The signal is recovered after rectification and filtering.

(ii) The second method utilises a phase sensitive demodulator so that both the tensile and compressive strains can be detected without initially unbalancing the bridge. Polarity of the output determines whether the strain is of tensile or compressive nature. In practice a large number of gauges are mounted on a test member, and the strain values are desired at about the same time. Therefore very fast switching of gauges is incorporated. This might lead to errors unless special care is exercised.

STRAIN GAUGE DATA REDUCTION

The aim of strain measurement is to obtain both the magnitude and direction of principal strains and hence of principal stresses. An arrangement of strain gauges, often called 'rosettes', is used for this purpose. Two most common rosettes are of star and delta types. These are shown in Fig. 5.14 (a) and (b) respectively. In practice the gauges are stacked

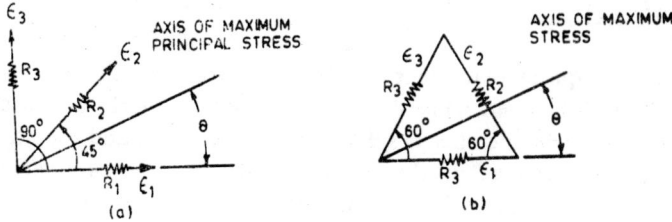

Fig. 5.14   Strain gauge data reduction

TABLE 5.2

Rosette Configuration→	Star	Delta
ϵ_{max}, ϵ_{min} (Principal strains)	$= \dfrac{\epsilon_1+\epsilon_2}{2} \pm \dfrac{1}{2}\Big[(\epsilon_1-\epsilon_2)^2 + (\epsilon_2-\epsilon_3)^2\Big]^{1/2}$	$= \dfrac{\epsilon_1+\epsilon_2+\epsilon_3}{3} \pm \dfrac{\sqrt{2}}{3}\Big[(\epsilon_1-\epsilon_2)^2 + (\epsilon_2-\epsilon_3)^2+(\epsilon_3-\epsilon_1)^2\Big]^{1/2}$
σ_{max}, σ_{min} (Principal stresses)	$= \dfrac{E(\epsilon_1+\epsilon_2)}{2(1-\mu)} \pm \dfrac{E}{\sqrt{2}\,(1+\mu)} \Big[(\epsilon_1-\epsilon_2)^2+(\epsilon_2-\epsilon_3)^2\Big]^{1/2}$	$= \dfrac{E(\epsilon_1+\epsilon_2+\epsilon_3)}{3(1-\mu)} \pm \dfrac{\sqrt{2}\,E}{3(1+\mu)}\Big[(\epsilon_1-\epsilon_2)^2+(\epsilon_2-\epsilon_3)^2 + (\epsilon_3-\epsilon_1)^2\Big]^{1/2}$
Orientation angle θ	$\tan 2\theta = \dfrac{2\epsilon_2-\epsilon_1-\epsilon_3}{\epsilon_2-\epsilon_3}$ θ is in the I quadrant if $\epsilon_2 < \dfrac{\epsilon_1+\epsilon_3}{2}$, θ is in the II quadrant if $\epsilon_2 > \dfrac{\epsilon_1+\epsilon_3}{2}$.	$\tan 2\theta = \dfrac{\sqrt{3}\,(\epsilon_3-\epsilon_2)}{2\epsilon_1-\epsilon_2-\epsilon_3}$ θ is in the I quadrant if $\epsilon_1 > \epsilon_2$ θ is in the II quadrant if $\epsilon_3 < \epsilon_2$.
γ_{max} (Maximum shear stress)	$= \dfrac{E}{\sqrt{2}\,(1+\mu)}\Big[(\epsilon_1-\epsilon_2)^2 + (\epsilon_2-\epsilon_3)^2\Big]^{1/2}$	$= \dfrac{\sqrt{2}}{3}\dfrac{E}{(1+\mu)}\Big[(\epsilon_1-\epsilon_2)^2 + (\epsilon_2-\epsilon_3)^2+(\epsilon_3-\epsilon_1)^2\Big]^{1/2}$

over each other.   Let the measured strains by the gauges designated R_1, R_2 and R_3 be ϵ_1, ϵ_2 and ϵ_3 respectively.   The principal strains, principal stresses, and their directions along with the maximum shear stress are expressed in terms of the measured values of strains for both the rosettes types and are presented in Table 5.2.

5.5   Grid technique

The method requires placement of reference marks over the area of interest on the test member.   The distances between the marks are measured before and after the member is subjected to stress field, and the strain is computed as the change in the length divided by the original length between the marks.   The distance between the two adjacent marks is the gauge length over which the strain is averaged.   The reference marks are in the form of a continuous grid pattern; rectangular, polar or of any other forms.   Sufficient information is made available from the measurements on these patterns to determine magnitudes and directions of principal stresses at every point.   For example, with a rectangular grid, normal strains in these directions can be determined by the measurement of the change in length of the sides and diagonal.   With the normal strains in three directions known, the principal strains and stresses can be determined by using the appropriate rosettes equation.

The grid pattern may be formed on the surface of the test member by any one of the several methods, viz.
   (i) by drawing or scribbling,
  (ii) by photographic printing,
 (iii) by photoetching, and
 (iv) by cementing a photographically prepared grid net work on the surface.
The coarse grids can be made by drawing or scribbling on the surface of the test member.   They find applications for the measurement of large strain.   Finer grid patterns are used where better accuracies are desired. These grids are photographically made either on the emulsion or photoetched directly on the surface.   Photo grids have been found to possess remarkable tenacity and have been applied to metal specimen which were later deep drawn.   The grid pattern made on a photographic sheet can also be bonded to the test member.

The necessary grid dimensions can be measured in a number of ways. In some cases, the measurements are made on the test member with a microscope having a micrometer eye-piece.   The method is indeed very laborious, requiring measurements of at least three dimensions on each grid.   Care must be taken to ensure that the magnification is the same when the readings are taken before and after the test member is subjected to stress field.   This is particularly so when high magnification of the microscope is used.   More often, photographs with a camera having a high resolution and distortion-free lens are taken as the test progresses and are evaluated leisurely. The effect of film shrinkage due to wet photographic process should be taken into account while evaluating the photographs.

When the measurements are to be made on a highly curved surface, a corner etc., Miller's replica technique can be used. The technique is very laborious requiring measurements at every grid square.

5.6 Moiré fringe Method

The grid technique requires measurements at every grid square for steep strain gradients, and thus is very laborious and time consuming. On the other hand, Moiré fringe technique provides full field details to the experimenter. The Moiré fringe effect is the formation of alternately bright and dark bands when one of the two patterns (usually line gratings) is placed on top of the other and either rotated or translated. For strain measurement, one of the gratings is mounted on the test member thus it undergoes the deformation identical to the test member. This grating is called a 'test' grating. The deformed grating is compared with a second grating, called the 'master' grating. The comparison is carried out over the whole field and hence Moiré technique is called a whole field method. One can use Moiré technique both for in-plane and out-of-plane deformation measurements.

The gratings used for Moiré work have equal dark and transparent regions, the width of the dark and transparent regions need not be equal but this gives maximum sensitivity. The Moiré gratings often used have a frequency of 40 lines/mm, sometimes upto 80 lines/mm gratings are also used. The frequency higher than this, when used, should incorporate diffraction phenomenon when analysing the formation of Moiré fringes.

In-Plane Deformation

The application of Moiré technique to the in-plane deformation measurement is dealt under the following three categories:

    (i)   Rigid body motion,

    (ii)   Deformation in one direction, and

    (iii) Rotation.

Rigid body motion

Consider two Moiré gratings of equal pitch (the frequency of the gratings is same: inverse of frequency is pitch) placed one over the other As one grating moves over the other, starting from a position when the two gratings are laid in such a way that the dark element of one falls over the bright element of the other so that there is no transmission, the transmission will increase over the whole field. If this two-grating arrangement is illuminated by a collimated light, and transmitted light is collected and fed to a photo sensor, the output of the photo sensor will be periodic with a period equal to the period of the grating. Therefore, by counting the periods of the output from a fixed datum the rigid body motion can be obtained easily. Initially the gratings are set parallel to each other and they remain so during the measurement.

Measurement of deformation along one axis

The grating on the test member is mounted in such a manner that the grating elements are perpendicular to the direction of deformation. Therefore, loading of the test member will result in the change of period of the test grating, which is compared with the master grating initially set parallel to the former. The experimental arrangement is shown in Fig. 5.15. Let the period of master grating be p and that of deformed grating be p'.

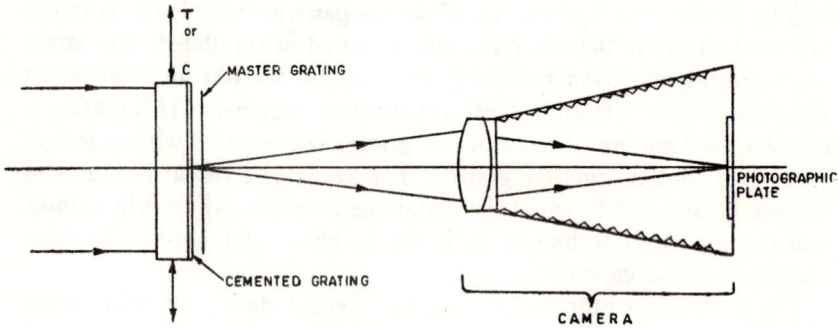

Fig 5.15   An arrangement to observe Moire fringes

Because of this mismatch, Moiré fringes will be formed; the distance d between two Moiré fringes will be governed by the deformation such that the total deformation from a reference which gives the position of a fringe, to the next fringe is one period of the grating. Let there be m grating periods of master grating that fall between two Moiré fringes, then

$$d = mp = (m \pm 1) p'$$

$$p' = \frac{m}{m \pm 1} p = \frac{d}{\dfrac{d}{p} \pm 1}$$

Writing $p' = p(1 \pm \epsilon)$, one obtains

$$\epsilon = \frac{p}{d - p} \text{ for tensile strain}$$

and

$$\epsilon = - \frac{p}{d + p} \text{ for compressive strain}$$

where ϵ is the strain.

The fringe direction in this case is parallel to the grating element and both compressive and tensile strains are calculated from the knowledge of p and the measurement of d. Usually $p \ll d$, and hence equal compressive and tensile strains produce identical fringe structure, and

$$\epsilon = \pm \frac{p}{d}.$$

The displacement normal to the grating elements on the specimen is therefore NP, where N is the order of Moiré fringe. Thus in order to determine the total displacement at any point, the number of Moiré fringes from a known datum is to be found out. For an arbitrary displacement, experiments are conducted on two models with gratings laid parallel and perpendicular. The Moiré fringes are loci of points having the same component of displacement in a direction perpendicular to the elements of the master gratings. Both the components of displacement are obtained from the two Moiré patterns obtained from orthogonally mounted gratings.

Pure rotation of two Moiré gratings

In this case consider two gratings of different frequencies; this situation is identical to the one where one of the two identical gratings has undergone a linear deformation. Later on the general result will be discussed for identical gratings. Let one of the gratings be with period b and elements parallel to x-axis, while the other with period a and elements inclined at an angle θ with the x-axis. The equations of these gratings are:

$$y = bh$$

and

$$y = x \tan \theta + \frac{ak}{\cos \theta}$$

where the integer constants h and k can take both negative and positive values, i.e, $0, \pm 1, \pm 2, \pm 3, \ldots$ and define various grating elements. The equation of the Moiré fringes will be the following indicial equation:

$$h - k = m$$

where m is again an integer taking $0, \pm 1, \pm 2, \ldots$ values and is a label to various Moiré fringes.

Therefore, the equation of Moiré fringes is

$$y = x \tan \phi - \frac{md}{\cos \phi}$$

where the orientation ϕ with x-axis and period d of the Moiré fringes are

$$\tan \phi = \frac{b \sin \theta}{- a + b \cos \theta}$$

and $\qquad d = \dfrac{ab}{(a^2 + b^2 - 2ab \cos \theta)^{1/2}}$.

For a special case, when one of the two identical gratings has suffered a pure rotation of θ

$$\phi = \theta/2$$

and $\qquad d = \dfrac{a}{2 \sin (\theta/2)} = \dfrac{p}{2 \sin (\theta/2)}$.

Thus the Moiré fringes are bisection of the two grating elements. For small angle

$$\theta = p/d$$

This geometrical approach for studying the Moiré formation is not restricted to linear gratings but can indeed be applied to any grating structure viz. circular, radial etc.

Since shear strain γ_{xy} results in rotation, Moiré fringes can be used for the shear strain measurement. If the rotation is α, then the shear strain $\gamma_{xy} = \tan^{-1} \alpha$.

If the rotation due to shear along the x-direction is $(\alpha/2)_x$, then it can be measured by having both gratings parallel to x-axis, then

$$\theta_x = (\alpha/2)_a = 2 \sin^{-1} \left(\frac{p}{2d_x} \right).$$

Similarly the rotation due to shear along y direction is $(\alpha/2)_y$ and is measured by having gratings parallel to y-axis. Then

$$\theta_y = (\alpha/2)_y = 2 \sin^{-1} \left(\frac{p}{2d_y} \right)$$

Therefore the shear strain γ_{xy} is given by

$$\gamma_{xy} = \tan^{-1} [(\alpha/2)_x + (\alpha/2)_y].$$

Determination of in-plane displacement

It has been mentioned earlier that both in-plane displacement components can be obtained from Moiré patterns of orthogonally laid gratings on two models. Assuming a cartesian reference system of axes x and y, the component of displacement parallel to x-axis is represented by u and the component parallel to y axis by v. Moiré fringes representing either u or v family of displacement are called isothetics. Isothetics are, therefore, loci of points having the same value of displacement component. Both the u and v isothetics can be obtained from the Moiré pattern. However it is instructive, at this moment to introduce the concept of displacement surface. The whole Moiré pattern can be visualised as a displacement surface where the height of a point on the surface above a reference plane represents the displacement of the point in a direction normal to the grating elements. In the case of a specimen having a uniform displacement, the displacement surface is an inclined plane. The slope of the surface at any point normal to the grating elements represents strain in that direction.

Consider now a situation when the Moiré pattern is obtained with grating having its elements parallel to the y-axis. Figure 5.16 shows a portion of the Moiré pattern. Two lines *ab* and *cd* are drawn parallel to

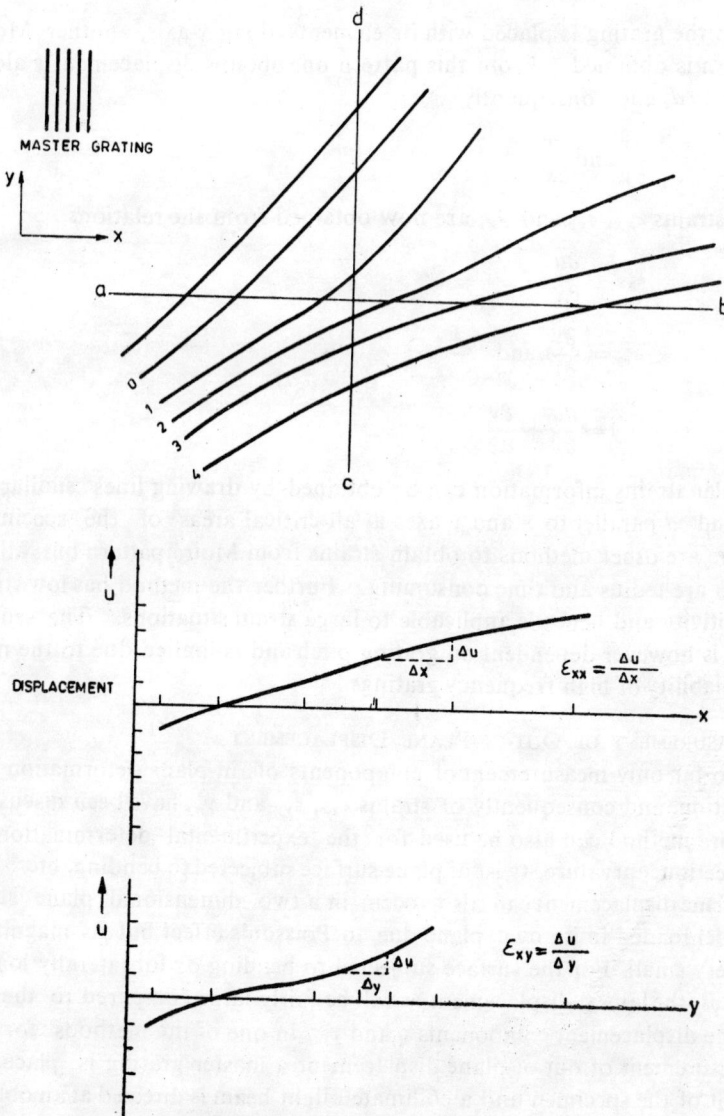

Fig. 5.16 *u*–Displacement curves from Moire pattern

the x and y axes respectively, and Moiré fringes are assigned arbitrary orders as only relative displacements are measured. The displacement u along x-axis is plotted against distance along ab and cd by noting that

$$\text{displacement } u = Np,$$

where N is the fringe order.

The tangents drawn to these curves at any point give the slopes

$$\frac{\partial u}{\partial x} \text{ and } \frac{\partial u}{\partial y}.$$

When the grating is placed with its elements along x-axis, another Moiré pattern is obtained. From this pattern one obtains displacement v along ab and cd, and consequently

$$\frac{\partial v}{\partial y} \text{ and } \frac{\partial v}{\partial x}.$$

The strains ϵ_{xx}, ϵ_{yy} and γ_{xy} are now obtained from the relations

$$\epsilon_{xx} = \frac{\partial u}{\partial x},$$

$$\epsilon_{yy} = \frac{\partial v}{\partial y}, \text{ and}$$

$$\gamma_{xy} = \frac{du}{\partial y} + \frac{\partial v}{\partial x}.$$

Similar strains information can be obtained by drawing lines similar to ab and cd parallel to x and y axes at all critical areas of the specimen. There are other methods to obtain strains from Moiré pattern but all of them are tedius and time consuming. Further the method has low strain sensitivity and hence is applicable to large strain situations. The sensitivity is however dependent on grating pitch and is limited due to the non-availability of high frequency gratings.

MEASUREMENT OF OUT-OF-PLANE DISPLACEMENT

So far only measurement of components of in-plane deformation and rotation, and consequently of strains ϵ_{xx}, ϵ_{yy} and γ_{xy} have been discussed. Moiré method can also be used for the experimental determination of deflection, curvature, twist of plane surface subjected to bending, etc. Out of plane displacement can also occur in a two dimensional plane stress model loaded in its own plane due to Poisson's effect but its magnitude is very small. For the surface subjected to bending or for laterally loaded plates, the lateral displacement w will be fairly large compared to the in-plane displacement components u and v. In one of the methods for the measurement of out of plane displacement, a master grating is placed in front of the specimen and a collimated light beam is directed at an oblique incidence. The shadow of the master grating projected on the specimen acts as a test grating. The observations are made normal to the surface of the master grating. The Moiré pattern is formed between the master grating and its shadow. If a point of zero out of plane displacement exists on the specimen, the master grating can be positioned as to allow the Moiré fringe to pass through this point. The out of plane displacement w is then given by

$$w = \frac{Np}{\tan \phi}$$

where N is the order of the Moiré fringe and ϕ is the angle of incidence.

The Moiré method has also been used for the measurement of slope directly. The Moiré method is a whole field method. The sensitivity of the method depends on the grating pitch and the usual gratings have frequency of 40 1/mm to 80 1/mm. Higher sensitivity can be obtained either by fringe shifting or fringe mismatch technique. The gratings are obtainable either as stripable photo-gratings or may be photo-etched directly on the specimen. The main advantage of Moiré method is quick appraisal of information and wide range of deformations which could be measured.

STRAIN-DISPLACEMENT RELATIONSHIP

There are two definitions of strains which are commonly used in experimental strain analysis. They are

$$\epsilon^L = \frac{L_f - L_i}{L_i}\bigg|_{L_i \to 0} \qquad \text{Lagrangian description of strain.}$$

$$\gamma^L = (\pi/2 - \xi_f)$$

and

$$\epsilon^E = \frac{L_f - L_i}{L_f}\bigg|_{L_f \to 0} \qquad \text{Eulerian description of strain.}$$

$$\gamma^E = (\xi_i - \pi/2)$$

where ϵ and γ are the direct and shear strains,

L_f and L_i are the final and initial lengths,

ξ_f final angle which was initially a right angle.

ξ_i initial angle which after deformation becomes a right angle.

The above equations indicate that the difference between the two definitions lies in the choice of base length or base angle. General strain-displacement relations under both descriptions of strain without any mathematical derivation are given here. A detailed analysis may be found in the book by Durelli and Parks. These relations are:

$$\epsilon_x^L = \left[1 + 2\frac{\partial u}{\partial x} + \left(\frac{\partial u}{\partial x}\right)^2 + \left(\frac{\partial v}{\partial x}\right)^2\right]^{\frac{1}{2}} - 1$$

$$\epsilon_y^L = \left[1 + 2\frac{\partial v}{\partial y} + \left(\frac{\partial v}{\partial y}\right)^2 + \left(\frac{\partial u}{\partial y}\right)^2\right]^{\frac{1}{2}} - 1$$

$$\gamma_{xy}^L = \sin^{-1} \frac{\frac{\partial u}{\partial y} + \frac{\partial v}{\partial x} + \left(\frac{\partial u}{\partial x}\right)\left(\frac{\partial u}{\partial y}\right) + \left(\frac{\partial v}{\partial x}\right)\left(\frac{\partial v}{\partial y}\right)}{(1 + \epsilon_x^L)(1 + \epsilon_y^L)}$$

and

$$\epsilon^E_x = 1 - \left[1 - 2\frac{\partial u}{\partial x} + \left(\frac{\partial u}{\partial x}\right)^2 + \left(\frac{\partial v}{\partial x}\right)^2\right]^{\frac{1}{2}}$$

$$\epsilon^E_y = 1 - \left[1 - 2\frac{\partial v}{\partial y} + \left(\frac{\partial v}{\partial y}\right)^2 + \left(\frac{\partial u}{\partial y}\right)^2\right]^{\frac{1}{2}}$$

$$\gamma^E_{xy} = \sin^{-1}\frac{\frac{\partial u}{\partial y} + \frac{\partial v}{\partial x} - \left(\frac{\partial u}{\partial x}\right)\left(\frac{\partial u}{\partial y}\right) - \left(\frac{\partial v}{\partial x}\right)\left(\frac{\partial v}{\partial y}\right)}{(1 - \epsilon^E_x)(1 - \epsilon^E_y)}$$

The terms like $\frac{\partial u}{\partial y}$, $\frac{\partial v}{\partial x}$ are known as cross derivatives. The relations under particular cases for example small rotations $\left[\left(\frac{\partial u}{\partial y}\right)^2 \rightarrow 0, \left(\frac{\partial v}{\partial x}\right)^2 \rightarrow 0\right]$, small strains, etc., may be obtained easily from the general relations. In case of Moiré fringe analysis, small strains are assumed and hence $\epsilon_x = \frac{\partial u}{\partial x}$ etc.

5.7   Interferometry

This is also a whole field method and is not often used because of its extremely high sensitivity. Further, it requires an optically flat test member, therefore, limited only to plane members. However, in special circumstances, for example, in photo-elastic studies where isopachics are obtained by interferometry, its use is desired. Therefore, the basic principle of interference as applied to displacement measurement is first described.

Consider Fig. 5.17 in which a beam of monochromatic light from a point source S is collimated by the lens L_1. This beam is divided into two beams which travel at right angles to each other. On reflection from a flat mirror M and optically worked test member T, the beams recombine at the beam splitter and are directed in the same direction. Lens L_2 is used to project interference fringes. Assuming that the flat mirror M and test surface T are orthogonal to each other, i.e., the reflected beams proceed exactly in the same direction; the interference field will then be uniformly illuminated. The interference condition is

$$2d = n\lambda, \, n = 0, \pm 1, \pm 2, \ldots$$

where $2d$ is the path difference between the interfering beams. When this condition is satisfied, the illumination of the interference field will be maximum. If the test member is translated by $\lambda/4$ normal to the optic axis, the field will have minimum illumination. Every $\lambda/2$ displacement will, however, restore the maximum illumination. If the translation L is an integral multiple of $\lambda/2$, then m fringe will pass when the specimen is translated by this distance i.e.

$$2(d + L) = (n + m)\lambda$$

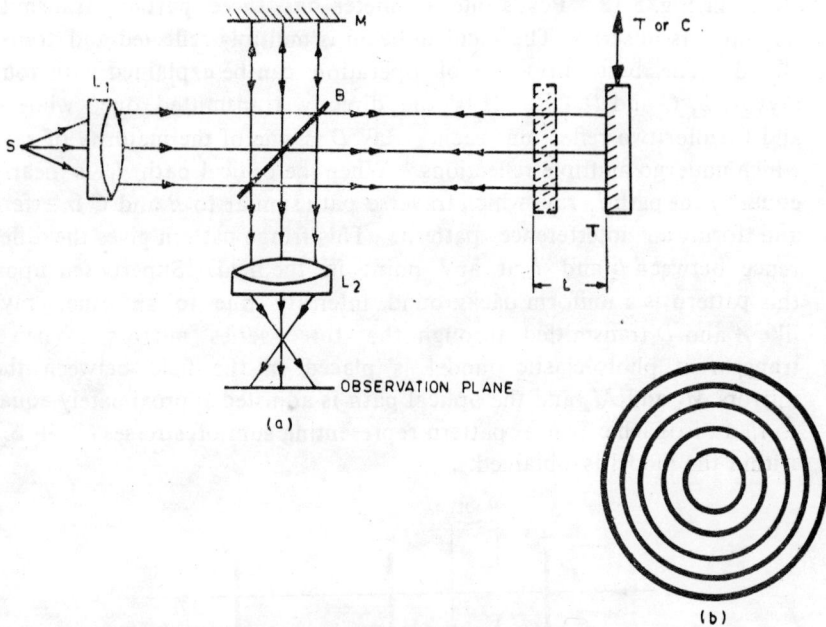

Fig. 5.17 (a) Interferometer for measurement of *L* and
(b) An interferogram

Therefore *L* can be measured, even to a fraction of wavelength, by counting the number of fringes. However, in a situation where the specimen undergoes non-uniform out of plane displacement, the interference fringes are loci of constant displacement. As an example consider an edge-clamped diaphragm which is uniformly loaded. The displacement is zero at the edge and maximum at the center. Circular fringes will be observed in the whole field, each representing a constant displacement of $\lambda/2$. Therefore the displacement profile can be obtained.

The interference technique is essentially sensitive to out-of-plane deformations. But by certain modifications the technique can be made sensitive to in-plane deformation also.

Due to its extremely high sensitivity, response to the out-of-plane deformation and restrictions on the surface quality of the specimen, the technique is not often used for deformation measurements.

For stress-analysis one needs to know the magnitude and direction of principal stresses S_1 and S_2. The difference of principal stresses $(S_1 - S_2)$ is obtained from the isochromatics in photo-elasticity. Therefore, one should either know S_1 or S_2 or $(S_1 + S_2)$ in order to separate the stresses. The sum of stresses can be obtained by the measurement of transverse deformation of the photo-elastic specimen when loaded. This can be easily measured with high sensitivity using interferometry. One very important and often used interferometer for this purpose is the Post's series interferometer. The schematic of the interferometer is shown in Fig. 5.18. Post's interferometer has three partially transmitting mirrors in series. The incident beam is multiply reflected and transmitted. The basic principle of operation can be explained with four rays A, B, C and D. Ray A is the directly transmitted one, while B and C suffer two reflections each. Ray D is one of the majority of rays which undergo multiple reflections. When the optical path l_1 is nearly equal to the path l_2, rays which traverse path similar to B and C interfere and form an interference pattern. This fringe pattern gives the difference between l_1 and l_2 at any point in the field. Superposed upon this pattern is a uniform background intensity due to all other rays like A and D transmitted through the three series mirrors. When a transparent photo-elastic model is placed in the field between the mirrors M_1 and M_2 and the optical path is adjusted approximately equal to l_1, an isopachic fringe pattern representing sum of stresses $(S_1 + S_2)$ within the model is obtained.

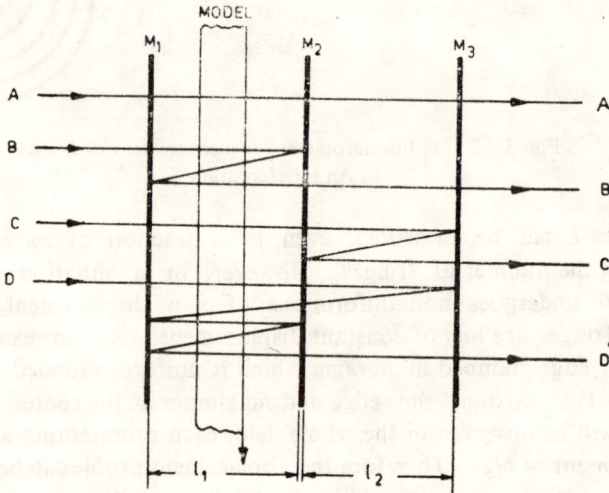

Fig. 5.18 Post's series interferometer

5.8 Photo-elasticity

This technique is based on the fact that a certain class of materials when subjected to stress become birefringent. The physical properties, here dielectric constant, become direction dependent. It is found that the directions of principal stresses are along the directions of principal refractive indices. Further the difference in the magnitude of principal stresses is proportional to the refractive index difference along these axes. That is,

$$(S_1 - S_2) = K(n_1 - n_2)$$

where n_1 and n_2 are refractive indices along the directions of principal stresses S_1 and S_2 and K is a constant. This relation holds good only within the elastic limit. Consider a plane-parallel slab of a transparent material subjected to stress. A beam of light will decompose into two linearly-orthogonally polarised beams as it enters the slab. These beams will propagate with different velocities, resulting in a net phase difference δ given by

$$\delta = \frac{2\pi}{\lambda} \, d \, (n_1 - n_2)$$

where d is the thickness of the slab and the incidence is assumed to be normal. These two beams will emerge in phase when $\delta = 2m\pi$ where $m = 0$, $\pm 1, \pm 2, \ldots$.

Therefore $(n_1 - n_2) = \dfrac{m\lambda}{d}$.

Thus $\qquad (S_1 - S_2) = mK \dfrac{\lambda}{d} = m \dfrac{F}{d} = mf$

where F is a constant called 'material fringe value' and f is another constant called 'model fringe value'.

It may be seen that F is a constant for the material and light used, and is independent of material thickness, while f depends on the material thickness as well. F is usually employed for comparing different photo-elastic materials, while f is convenient for stress conversion in individual tests. The relation which connects $(S_1 - S_2)$ with the fringe order is called as stress optic law and is the basis of photo-elasticity. The photo-elasticity is an experimental method of stress-analysis; it is an experimental method in a sense that an optical fringe pattern is experimentally observed but the state of stress in the member is to be theoretically related with this method. This is a full field method and capable of providing a very high accuracy. As has been said earlier this method provides an information regarding difference of principal stresses. Therefore, to affect separation of stresses, a knowledge of S_1 or S_2 or $(S_1 + S_2)$ is necessary. A method to measure the sum of stresses is described later.

There are two main kinds of arrangements used in photoelasticity: plane polariscope and circular polariscope.

PLANE POLARISCOPE

The functional elements of a standard plane polariscope and the transformations which the light undergoes when it passes through the plane polariscope are shown in Fig. 5.19 (a). The physical elements are:

    (i) a light source,
    (ii) a pair of polariser,
    (iii) a model, and
    (iv) an observation screen or a camera.

The lenses are used for collimating the beam and projecting the fringes at the screen. The model is placed in between the polariser and analyser. Let us assume that the beam emerging from the polariser is vertically

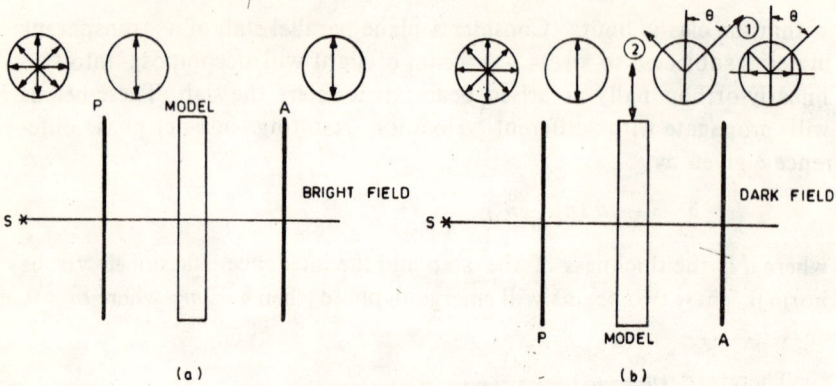

Fig. 5.19   Linear polariscope

polarised (i.e. E vector vibrating vertically). This is represented as $E = E_0 e^{i\omega t}$. If the analyser (other polariser) is crossed, i.e., its direction of transmission is perpendicular to that of the polariser, the field of view will be dark. On the other hand, if it is parallel, the field will be bright. At any other orientation, the intensity transmitted will follow $\cos^2 \theta$ law, where θ is the angle between the transmission axes of polariser and analyser. Figure 5.19 (b) shows the same arrangement with model stressed. Let one of the principal axes be tilted at an angle θ with the vertical. The amplitudes of light transmitted along the principal axes (1) and (2) are given by

$$E_1 = E_0 \cos \theta \, e^{i\omega t},$$

and

$$E_2 = E_0 \sin \theta \, e^{i(\omega t + \delta)}$$

where δ the phase difference introduced between the two beams and is given by

$$\delta = \frac{2\pi}{\lambda} \, d \, (n_1 - n_2)$$

n_1 and n_2 are principal refractive indices.

Only that component of the field, which is along the transmission direction of the analyser is allowed by the analyser. Assuming the analyser axis to be perpendicular with that of the polariser, the amplitude of the light transmitted is

$$= E_0 \cos \theta \sin \theta \, e^{i\omega t} - E_0 \sin \theta \cos \theta \, e^{i(\omega t + \delta)}$$

$$= \frac{E_0}{2} \sin 2\theta \, e^{i\omega t} \, [1 - e^{i\delta}]$$

The intensity transmitted is therefore given by

$$I(t) = I_0 \sin^2 2\theta \sin^2 (\delta/2)$$

An equation similar to this can also be derived when the analyser axis is parallel to the axis of the polariser.

The intensity in the transmitted beam becomes zero under two conditions:

Condition 1 (Isoclinics)

When $\theta = \dfrac{m\pi}{2}$; where $m = 0, 1, 2, \ldots$

The directions of principal stresses are parallel to the transmission axes of either the polariser or the analyser. So at all those points in the member where the principal stress directions are coincident with the transmission axes of the polariser and the analyser, there will be no transmission and these points will appear dark. Usually these points are on a continuous line. This line goes through all points where the principal stress has the same inclination, and is called isoclinic. The angle θ made by the transmission axes of the polariser and the analyser with the reference direction (usually vertical or horizontal) is called a parameter of the concerned isoclinic. By changing θ, isoclinics of different parameters are obtained. Further it should be noted that isoclinics are independent of wavelength of light or/and of retardation (stress). Usually a white light source with ,a convenient stress is used to find isoclinics.

Condition 2 (Isochromatics)

The intensity of transmitted light will be zero for any value of θ, provided

$$\delta = 2\,m\pi \text{ where } m = 0, \pm 1, \pm 2, \ldots,$$

or $\qquad d(n_1 - n_2) = m\lambda$

Using stress optic law,

$$m = \frac{1}{f}(S_1 - S_2)$$

Therefore, in the stressed member, wherever $(S_1 - S_2)$ is such that m takes 0 or integer values, the transmitted intensity will be zero. In monochromatic light, it corresponds to continuous dark fringes. However, in white light, the appearance will be coloured. Each band corresponds to a certain value of $(S_1 - S_2)$ and, hence, is called isochromatic. In monochromatic light, the fringes corresponding to $m = 0, 1, 2, \ldots$ are called fringes of zero, first, second order respectively.

CIRCULAR POLARISCOPE

Both isoclinics and isochromatics appear together in the plane polariscope, the isoclinics are broad, dark and hide the isochromatics. Therefore, elimination of former is often desirable. This is achieved in circular polariscopes. In short, plane polariscope is used to obtain direction of principal stresses, while circular polariscope to measure the difference of principal stresses. Figure 5.20 shows a schematic diagram of a circular polariscope. It consists of a polariser, an analyser and a

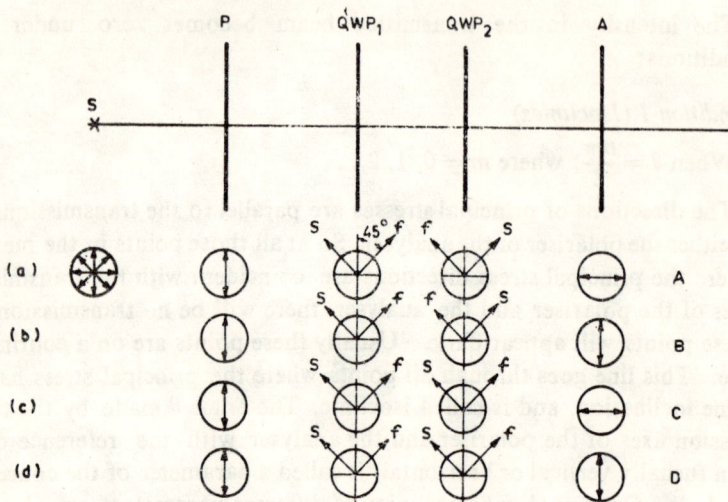

Fig. 5.20 Various arrangements in a circular polariscope

pair of quarter wave plates. For the purpose of analysis, a configuration has been chosen in which the polariser and the analyser are crossed with polariser's transmission axis vertical and the fast and slow axes of quarter wave plates QWP_1 and QWP_2 orthogonal to each other and inclined at an angle of 45° with the transmission direction of the polariser. The transfer of polarisation states through various physical elements is described below. The electric field after the polariser is $E = E_0 e^{i\omega t}$. This is also the field incident on the quarter wave plate. Let the fields transmitted along fast and slow axes of QWP_1 be E_1 and E_2 respectively, then

$$E_1 = \frac{E_0}{\sqrt{2}} e^{i\omega t}$$

$$E_2 = \frac{E_0}{\sqrt{2}} e^{i(\omega t + \pi/2)}$$

These fields are incident on the quarter wave plate QWP_2 with axes orthogonal to that of QWP_1. Therefore, the fields E_3 and E_4 transmitted along slow and fast axes are

$$E_3 = \frac{E_0}{\sqrt{2}} e^{i(\omega t + \pi/2)},$$

$$E_4 = \frac{E_0}{\sqrt{2}} e^{i(\omega t + \pi/2)}$$

The fields transmitted by the analyser are those components which are parallel to the transmission axis of the analyser. Therefore,

$$E_{3t} = \frac{E_0}{2} e^{i(\omega t + \pi/2)},$$

$$E_{4t} = \frac{E_0}{2} e^{i(\omega t + \pi/2)}$$

The net amplitude transmitted by the analyser is a vector sum of E_{3t} and E_{4t} resulting in

$$E_t = E_{3t} - E_{4t} = 0$$

Thus, there is no intensity transmitted, resulting in a dark field. This is one of the possible arrangements. Given below are all the four arrangements:

S. No.	Arrangement	Quarter wave plates	Analyser and Polariser	Field
1.	A	Crossed	Crossed	Dark
2.	B	Crossed	Parallel	Bright
3.	C	Parallel	Crossed	Bright
4.	D	Parallel	Parallel	Dark

Figures 5.20 (a) to (d) show the functions of various elements in the above arrangement. However, the necessary modifications can be introduced when the stressed member is placed between the two quarter wave

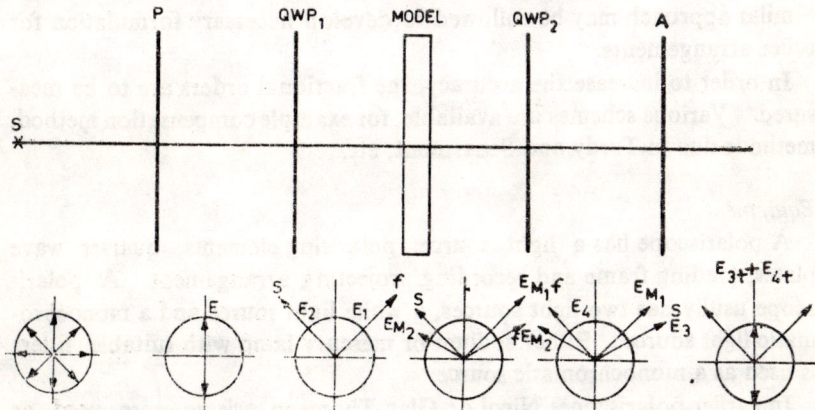

Fig. 5.21 Circular polariscope in arrangement B with model stressed

plates. Considering arrangement B and the principal axes of stressed member inclined at an angle θ with the axes of quarter wave plate as shown in Fig. 5.21, the field transmitted from the member along its axes is

$$E_{m1} = (E_1 \cos \theta + E_2 \sin \theta)\, e^{i\omega t}$$

$$E_{m2} = (E_2 \cos \theta - E_1 \sin \theta)\, e^{i(\omega t + \delta)}$$

or $\quad E_{m1} = \dfrac{E_0}{\sqrt{2}}\, e^{i(\omega t + \theta)}$

$$E_{m_2} = \frac{E_0}{\sqrt{2}}\, e^{i(\omega t + \theta + \pi/2 + \delta)}$$

where $\delta = \frac{2\pi}{\lambda}\, d(n_1 - n_2)$.

The field transmitted by the plate QWP_2 can be obtained by taking the components of E_{m1} and E_{m2} along the axes of the quarter wave plate. Further in the B arrangement, the fields E_{3t} and E_{4t} which are the components of E_3 and E_4 along the transmission axes of analyser will be added, resulting in the net field E_t where

$$E_t = E_{3t} + E_{4t} = \frac{E_0}{2}\, e^{i(\omega t + \pi/2)}\,(1 + e^{i\delta}).$$

Hence the intensity transmitted will be

$$It = I_0 \cos^2(\delta/2)$$

The maxima of intensity occur when

$$\delta = 0,\, 2\pi,\, 4\pi \ldots$$

and the minima of intensity occur when

$$\delta = \pi,\, 3\pi,\, 5\pi$$

Similar approach may be followed to develop necessary formulation for other arrangements.

In order to increase the accuracy, the fractional orders are to be measured. Various schemes are available, for example compensation method, methods due to Tardy and Senarmont, etc.

Equipment

A polariscope has a light source, polarising elements, quarter wave plates, loading frame and recording/projecting arrangement. A polariscope usully has two light sources, a white light source and a monochromatic light source. Either sodium or mercury lamp with suitable filters is used as a monochromatic source.

In earlier polariscopes Nicol or Glan-Thomson prisms were used as polariser and analyser; the aperture thus was very much limited and hence projection arrangements were incorporated. Now the polariscopes use polaroid sheets, which are available in very large sizes. The polaroids are not very good polarisers but are quite adequate for photoelastic work.

The quarter wave plates are usually thin sheets of mica with thickness such as to intrduce a path change of $\lambda/4$. Therefore, they are wavelength sensitive. However, now quarter wave plates which are almost achromatic are available as plastic sheets; the stress birefringence is made use of for necessary phase change.

Loading frame is an element which is designed by the investigator in consideration with the shape, size, etc. of the model and loading capacity. Invariably dead weight or dead weight magnified by a lever is adopted in the design.

For projection, a good quality lens is used or a permanent record can be made by focussing the camera at the fringe plane.

Materials

A number of materials for models have been used; they include several types of glasses, celluloid, gelatin, rubber, cellulose nitrate, vinyls, casto-lite, phenol formaldehydes, kriston, CR–39, catalin 61–893, polyesters, epoxies and urethane. Of these most commonly employed are epoxy resin, castolite, catalin 61–893, Columbia resin (CR–39) and urethane. The model made of these materials can be loaded and stress field studied as a function of loads or the stresses in the model may be frozen. The frozen stress field can be studied leisurely, or for three-dimensional photo-elasticity, the model is sliced and studied.

Data Analysis

It has been shown that, using polariscope, the directions and difference of principal stresses can be experimentally determined. In order to deter-mine the complete state of stress at any point, one should know indivi-dual stresses S_1 and S_2. Therefore, an experimental measurement of S_1 or S_2 or $(S_1 + S_2)$ is carried out by some alternative methods. There are a number of schemes available for this. However, only a few schemes are discussed here. It is known that the strain component perpendicular to the surface of the slab is given by

$$\epsilon_3 = \frac{S_3}{E} - \frac{\mu}{E}(S_1 + S_2)$$

where S_3 is the stress component in that direction, and μ and E are the Poisson's ratio and Young's modulus of the material of test member. Since $S_3 = 0$,

$$\epsilon_3 = -\frac{\mu}{E}(S_1 + S_2)$$

or

$$(S_1 + S_2) = -\frac{E}{\mu}\epsilon_3 = -\frac{E}{\mu}\frac{\Delta d}{d}$$

where Δd are the variations in the thickness which can be measured at various points, say by a sensitive dial gauge. This method is very cum-bersome as it requires measurement at large number of points. Any inter-ferometric method, especially the one suggested by Post, is a very useful one as it gives information about the whole field at a time. Its accuracy can be increased many times by placing the specimen between two high reflecting surfaces and only those beams which have suffered same number of reflections are allowed to interfere. This is known as fringe multiplication. Therefore, the stresses S_1 and S_2 are obtained from the following two equations:

$$S_1 - S_2 = m\frac{F}{d} = mf$$

and

$$S_1 + S_2 = -\frac{E}{\mu} \cdot \frac{\Delta d}{d}.$$

From these two equations, the state of stress at any point can be found out easily.

In another method, the model is rotated about one of the principal stress vectors at some point of interest so that the angle of incidence is ϕ. Then

$$S_1 - S_2 \cos^2\phi = m'\frac{F}{d} = m'f$$

where m' is the order of fringe at the point determined with oblique incidence. Therefore

$$S_2 = f \cdot \frac{m - m' \cos\phi}{(\cos^2\phi - 1)}$$

and

$$S_1 = f \cdot \frac{m \cos^2\phi - m' \cos\phi}{(\cos^2\phi - 1)}$$

Other methods of separation of stresses are due to Filon and Frocht. Electrical analogy may also be used.

It is not always necessary to make a model of the specimen for photo-elastic measurement. The specimen as such can be used, provided a layer of birefringent material is applied to it and a polariscope is used in reflection mode. The sensitivity of the method is relatively low due to the thin layer of birefringent material.

Exercises

1. A resistance strain gauge having $R = 120\Omega$ and $G_F = 2\ 0$ is placed in an equal-arm bridge in which all the resistances are each 120Ω. The bridge is excited by a d.c. source of 4.0 volts. Assuming that the galvanometer has a resistance of 100Ω, calculate the detector current in microamperes per μm/m of strain.

2. In a constant d.c. voltage excited bridge two strain gauges R_1 and R_1' are placed in series in arm 1 and are subjected to the same strain. The other arms of the bridge have resistance R_2, R_3 and R_4. Calculate the increase in bridge sensitivity over that obtained with a single strain gauge R_1.

3. In a d. c. excited bridge, strain gauges R_1 and R_4 are active and dummy gauges respectively and are of 120Ω each. The R_2 and R_3 are resistances, also of 120Ω each. The maximum gauge current is 25 ma. Calculate the maximum permissible value of the excitation voltage.

4. For the bridge in problem (3),
   (a) If the active gauge is bonded on to a steel member ($E = 2 \times 10^{11}$ N/m²) what is the bridge out put voltage per 10^7 N/m² of stress ?

(b) What bridge output would be developed by bonding the active gauge on to a steel member, in the absence of temperature compensation, whose temperature is increased by 40°C? What stress-value would be represented by this voltage ?

(c) What is the value of the calibration resistor that would give an output equal to that obtained by subjecting the steel member to 10^8 N/m² stress ?

Given: Thermal expansion coefficients of steel and material of the gauges are 12×10^{-6} and 25×10^{-6} respectively, temperature coefficient of resistance of gauge 10×10^{-6} Ω/Ω°C. Gauge factor is 2 and is independent of temperature.

5. An aluminium bar is loaded with a constant moment section as shown in Fig. 5.22. The deflection is measured with a dial gauge, and the strain is monitored by a strain gauge mounted on the center of the bar. The gauge resistance is measured as a function of deflection with a resistance measuring bridge. From the data given, determine the gauge factor of the gauge.

$d = 6$ mm, $w = 25$ mm, $S = 300$ mm, $X = 200$ mm, $Y = 150$ mm.

Gauge resistance ohms	121.3	120.7	120.2	119.5
Dial gauge reading mm	0	0.10	0.20	0.30

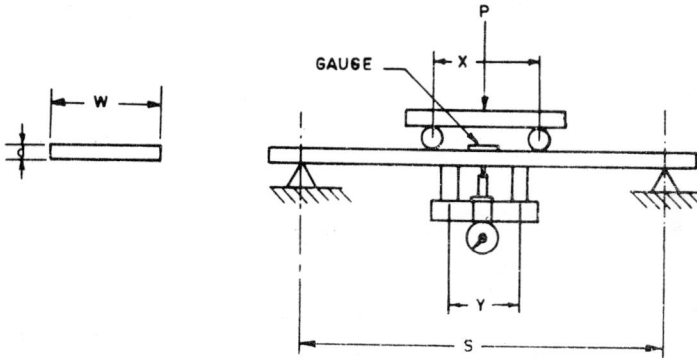

Fig. 5.22

6. A delta rosette is mounted on a steel ($E = 2 \times 10^{11}$ N/m²) plate which is subjected to load. The following strain values are indicated:

$\varepsilon_1 = 395$ μm/m, $\varepsilon_2 = 80$ μm/m, and $\varepsilon_3 = -250$ μm/m.

Calculate the principal strains and stresses, the maximum shear stress, and the orientation angle for the maximum principal stress.

7. A rectangular rosette is mounted on a steel plate which is subjected to load. The following strains are indicated :

$\varepsilon_1 = 560$ μm/m, $\varepsilon_2 = 150$ μm/m, and $\varepsilon_3 = -475$ μm/m.

Calculate the principal strains and stresses, the maximum shear stress, and the orientation angle of the maximum principal stress.

8 A Moiré grating of 8 lines/mm pitch is cemented to a slab of urethane rubber having thickness and width of 12 mm and 125 mm respectively. A compressive load of 10 Kgf is applied to the slab. Calculate the pitch of the Moiré fringes.

Given: E for urethane rubber is 41×10^5 N/m^2 and its Poisson's ratio μ is 0.474.

9. Two gratings of pitches of 8.0 lines/mm and 8.2 lines/mm respectively are inclined at an angle of 30° to each other. Calculate the pitch of Moiré fringes. What would be the pitch of Moiré fringes if the angle between them is increased to 45°?

10. A quarter wave plate is placed between a polariser and an analyser with its axis at 45° with respect to the transmission axis of the polariser. Show that the amount of light passing through the analyser is independent of its orientation.

11. A photo-elastic model is loaded with a constant moment section as shown in Fig. 5.23(a). If the dark field is used, the fringe pattern in the constant moment section appears as shown in Fig. 5.23(b). The moment in this section is 40 Kgf-cm. What is the material fringe constant (F) of this material?

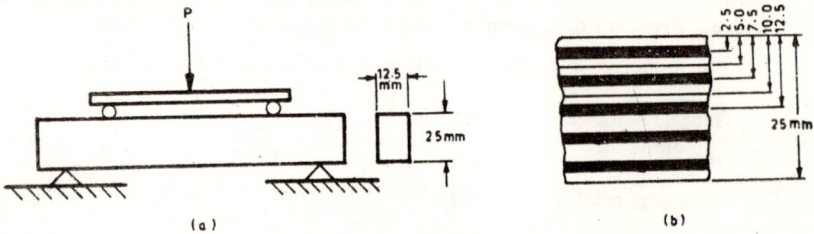

Fig. 5.23

Measurement of Pressure

6.1 Introduction

Pressure is described as force per unit area and is analogous to stress. Normally, it is referred to fluids only and arises due to exchange of momentum between the molecules of the fluid and walls of the container. The total exchange depends upon the total number of molecules striking the wall per unit time and the average velocity of the molecules.' For perfect gases, using kinetic theory, it can be shown that the pressure imparted by it at the walls of the container is given by

$$P = \frac{1}{3} nmv_{rms}^2$$

where n is the molecular density, m the mass of the molecule and v_{rms} is its root mean square velocity. Further, it can be shown that

$$v_{rms} = \sqrt{\frac{3kT}{m}}$$

where k is the Boltzmann constant related to the Avogadro's number N by $R = Nk$, R is the gas constant and T is the absolute temperature of of the gas. Therefore, the pressure is given by

$$p = nkT$$

with $kT = (1/2) mv_{rms}^2$. So the very concept of pressure arises solely due to the kinetic nature of gas molecules. This interpretation of pressure of a perfect gas is due to kinetic theory.

The description of pressure imparted by liquids is not so simple. The kinetic theory of liquids is well developed where the interaction of molecules among themselves has been taken into account. It must, however, be noted that the pressure imparted by liquid molecules due to their momentum exchange is not a complete description; the liquid head is an additional pressure imparting process. It is somewhat difficult to speak of the mechanism of pressure transmission at very low pressures of the order of 10^{-9} mm or lower, whereas the average size of the gas molecule is only 10^{-7} mm.

However, in practice, pressure is considered as an average physical quantity. In quite a few methods of the pressure measurements, force is measured and pressure is calculated from this by utilising the value of effective area. In other methods, indirect ways are used to obtain pressure.

6.2 Definition of pressure terms

Absolute Pressure: Absolute pressure refers to the absolute value of force per unit area exerted by a fluid on the walls of its container. It is the fluid pressure measured above a perfect vacuum.

Atmospheric Pressure: The pressure exerted by the earth's atmosphere, as commonly measured by a barometer. At sea level, its value is close to 1.013×10^5 N/m² absolute, decreasing with altitude.

Gauge Pressure: Gauge pressure indicates the difference between the absolute pressure and local atmospheric pressure.

Differential Pressure: This is the difference between two measured pressures.

Vacuum: Vacuum represents the amount by which the atmosphere pressure exceeds the absolute pressure.

These definitions are illustrated in Fig. 6.1. The pressure at a point in a fluid is equivalent to the weight of a certain column of the fluid acting on unit area at that point, i.e. $p = \rho g h$. This h is termed as the head.

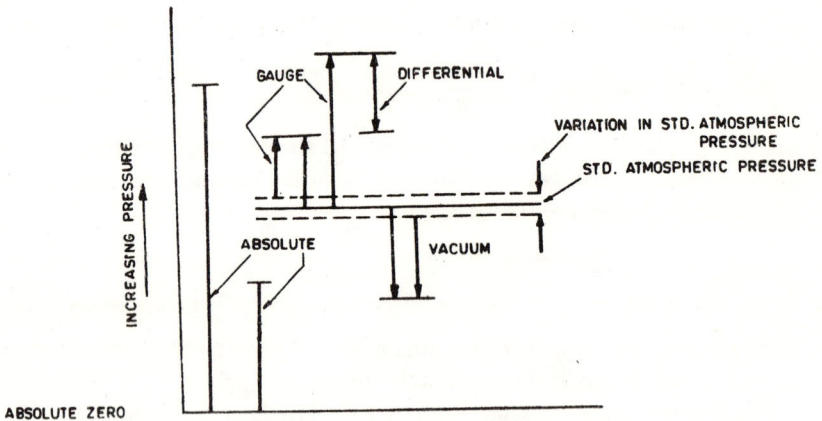

Fig. 6.1   Definition of pressure terms

6.3 Units of pressure, conversion factors and standards

The pressure is measured either as force per unit area or a head of a liquid column. It is, therefore, expressed as N/m², or in mm of water or mercury. The relations between the different units are listed in Table 6.1.

TABLE 6.1 Units of pressure and their conversion

Units	1 psi	1 Kgf/cm²	1 microbar = 1 dyne/cm²	1 N/m² = 1 Pa	1 mm Hg	1 mm w.c.
* 1 psi	1	7.03×10^{-2}	6.895×10^{4}	6.895×10^{3}	57.716	703.087
* 1 Kgf/cm²	14.223	1	98.067×10^{4}	98.067×10^{3}	735.56	10^{4}
* 1 dyne/cm² = microbar	14.50×10^{-6}	1.02×10^{-6}	1	0.1	7.5×10^{-4}	101.971×10^{-4}
* 1 Pa (= 1N/m²)	1.45×10^{-4}	10.197×10^{-6}	10	1	7.5×10^{-3}	10.2×10^{-2}
* 1 mm Hg or torr	1.934×10^{-4}	13.595×10^{-4}	1333.24	133.32	1	13.595
* 1 mm w.c.	14.22×10^{-4}	10^{-4}	98.067	9.807	7.36×10^{-3}	1

Some of the terms used in the text are
psia=pounds per square inch absolute
psig=pounds per square inch gauge
Pa = N/m² = Pascal
Standard atmospheric pressure = 760 mm Hg
= 14.696 psia
= 1.013×10^{5} N/m²
= 1.013×10^{5} Pa

Pressure Standards

Pressure standards are the basis for all pressure measurements. A gauge is always calibrated against a pressure standard and many different types are available, depending on the pressure and accuracy required. Typical standards include:

- (i) 'dead weight testers' which provide a known force by means of standard weights;
- (ii) 'liquid column testers' which compare the tested device against a known liquid head, and
- (iii) high accuracy 'test gauges' which themselves must be periodically checked against a more accurate standard.

6.4   Methods of measuring pressure

Pressure measuring techniques can be broadly classified in the following three groups:

- (i) Balancing the pressure exerted by fluid (usually mercury) column like in manometers, McLeod gauge etc.
- (ii) Measurement of elastic deformations of elements like membrane, diaphragm, Bourdon tube, etc.
- (iii) Measurement of electrical quantities like in Pirani and Penning gauges, Bridgman gauge, etc.

The ranges of pressure attained and the various instruments (gauges) used to measure pressure are shown in Table 6.2. The working principles of some of the common pressure measuring instruments will be discussed in this chapter.

6.5   Dead weight gauge tester

A schematic of the tester is shown in Fig. 6.2. It is a set up for the calibration of a pressure gauge. The tester consists of an accurately machined piston which is inserted into a close fitting cylinder. The cross-sectional areas of both the piston and cylinder are accurately known. The chamber and the cylinder of the tester are filled with a clean oil. One end of the piston carries a number of masses of known weights. Fluid pressure is applied to the other end of the piston by a fluid pump. When sufficient pressure is reached, piston-weight combination freely floats. When constrained to float freely within limit stops, the fluid pressure as indicated on the gauge must be equal to the weight of the weight-piston combination divided by the effective area of the piston, i.e.

$$P_{DW} = F_E/A_E$$

TABLE 6.2 Summary of some instruments for pressure measurement

Type		Name	Range (mm Hg)	Uncertainty
(i)	(a)	Dead weight testers	0 to 10^6	0.01 to 0.05% of the reading
	(b)	Manometers	10^{-1} to 10^4	0.02 to 0.2% of the reading
	(c)	Micromanometers	3×10^{-3} to 3	1% of the reading upto 10^{-3} mm Hg
	(d)	McLeod gauge	1 to 10^{-6} (absolute)	1% of the reading
(ii)	(a)	Bourdon tubes	10^2 to 10^6: upto 10^7 with certain types	upto 0.5% of full scale
	(b)	Bellows	10 to 10^4	1% of the reading
	(c)	Diaphragms (Mechanical)	0.5×10^3 to 10^5	about 0.5% of the reading
	(d)	Diaphragms (Strain gauge bonded)	0 to 10^6	0.25% or 'better' of the reading
(iii)	(a)	Resistance gauge (Bridgman gauge)	10^4 to 10^9	$\pm$ 1% of the range for manganin transducer
	(b)	Conductivity gauge (Pirani gauge)	10^{-4} to 0	
	(c)	Piezo-electric gauge	10^2 to 10^6 dynamic	
	(d)	Ionisation gauge	10^{-3} to 10^{-8} the range can be extended down to 10^{-12}	
	(e)	Radiometric gauge (Knudsen gauge)	10^{-3} to 10^{-6} (Absolute)	

where F_E the equivalent force of piston-mass combination depending on local gravity, air buoyancy, etc. while A_E is the effective area of the piston–cylinder combination and P_{DW} is the dead weight pressure. The effective area A_E depends on such factors as piston-cylinder clearance, pressure level, and temperature. The effective area is generally taken as mean of the cylinder and piston areas, but temperature affects this dimension. The effective area increases between 24 to 32 ppm/°C for

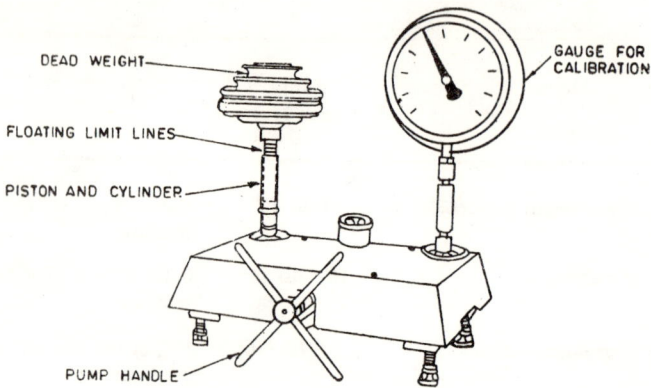

Fig. 6.2   Dead weight gauge calibrator

commonly used materials, and a suitable correction can be applied for this.

There will be fluid leakage out of the system through the piston-cylinder clearance. This fluid flow provides necessary lubrication between the two surfaces and reduces the friction. The piston is either oscillated or rotated to reduce the friction further. Due to the leakage of fluid, the system pressure must be trimmed continuously to keep the piston-weight combination floating within limit stops.

For highly accurate results, the following two corrections are to be applied:

Gravity Correction: The weights are usually given in terms of standard gravity value of 9.807 m/s² at sea level. Whenever gravity differs from this value due to altitude or latitude difference the equivalent force F_E would require a correction. The gravity correction term is given by

$$C_g = -\left(\frac{g_{local}}{g_{standard}} - 1\right)$$
$$= -[2.637 \times 10^{-3} \cos 2\phi + 32 \times 10^{-8}h + 5 \times 10^{-5}]$$

where ϕ is the latitude in degrees and h is the altitude above sea level in meters.

Buoyancy Correction: The weight and piston-combination displaces air, resulting in a buoyant force acting upward. The correction term for this is

$$C_b = -\left(\frac{W_{air}}{W_{weights}}\right)$$

where W_{air} and $W_{weights}$ are specific weights of air and piston weight combinations, respectively.

The corrected dead weight gauge pressure is then given by

$$P_{DW} = P_I (1 + C_g + C_b)$$

where P_I is the indicated pressure. The dead weight gauges usually are not capable of measuring pressures lower than the tare pressure (piston weight/effective area). However, this difficulty is overcome, in some such testers, by tilting piston-cylinder combination from vertical through an accurately known angle, thus giving full range from 0 Kgf/cm² to tare pressure.

6.6 Manometers

The atmospheric pressure at NTP is 760 mm of mercury (760 mm Hg). Therefore, the pressure can be measured by measuring the height of a mercury column. Usually mercury and water are preferred as manometer fluids because detailed information is available on their specific weights. Other low vapour-pressure and low density liquids like silicon oil are also used as manometer fluids.

The most common manometer is of U-tube type, partially filled with a suitable liquid. This is widely used for measurement of fluid pressure under steady state conditions. Usually capillary effects are neglected during measurements. To one end of the manometer is applied a reference pressure, say atmospheric pressure, while the other end is connected to the pressure to be measured. The measurement of fluid pressure requires that the fluid must have smaller density and be immiscible with manometer fluid. Consider a very general situation as shown in Fig. 6.3. The equality of pressure is taken at the level AB. Therefore,

$$P + W_P(h_1 + h_2 + \Delta h) = P_R + W_R h_2 + W_M \Delta h$$

$$\text{or } P - P_R = W_M \Delta h \left[1 + \frac{W_R h_2}{W_M \Delta h} - \frac{W_P}{W_M} \left(1 + \frac{h_1 + h_2}{\Delta h} \right) \right]$$

$$= W_M \Delta h C_h$$

where P and P_R are the unknown and reference pressures, W_P, W_R, and W_M are the specific weights (corrected for temperature and gravity) of fluids through which unknown and reference pressures are transmitted and of manometer fluid respectively, and Δh, h_1, and h_2 are the heights as shown in Fig. 6.3. The correction factor C_h is

$$C_h = \left[1 + \frac{W_R h_2}{W_M \Delta h} - \frac{W_P}{W_M} \left(1 + \frac{h_1 + h_2}{\Delta h} \right) \right]$$

If $W_P = W_R$, i.e., similar fluids are used to transmit pressures, then

$$C_h = 1 - \frac{W_P}{W_M} \left(1 + \frac{h_1}{\Delta h} \right).$$

In addition to this, if $h_1 = 0$, then

$$C_h = 1 - (W_P/W_M).$$

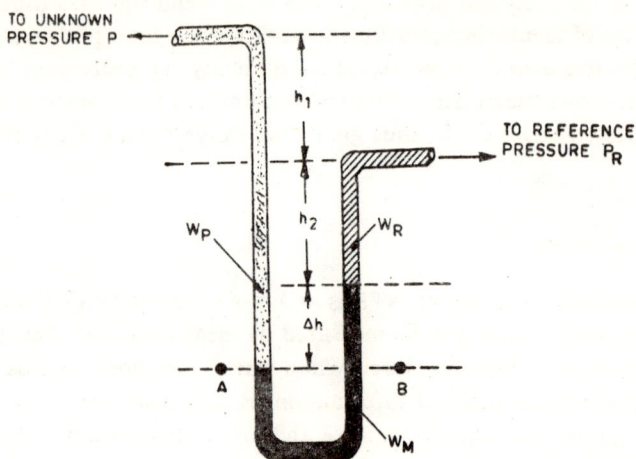

Fig. 6.3   Generalised manometer

In usual practice air is used for transmitting pressure, and hence C_h may be taken as unity, giving

$$P - P_R = W_M \Delta h,$$

which is a familiar expression.

It should be noted that Δh is measured parallel to the gravitational force and accuracy of pressure measurement depends on the accuracy with which Δh can be measured. Manometers can be used for the measurement of pressures at the moderate range of 10^{-1} mm to 10^4 mm Hg. Measurement of pressures higher than 10^4 mm Hg needs large manometers which are difficult to handle, while the measurement of low pressures is difficult due to very little difference in heights. The low density liquids can be used to increase the sensitivity; the use of inclined manometers is quite common.

In order to make very accurate measurements, the temperature, gravity and capillary corrections are made. The capillary effects are considerably reduced when tubes of diameter greater than 20 mm are used.

6.7   Micromanometers

The capabilities of U-tube manometers are extended by various types of micromanometers which serve as pressure standards in the range of 0.005 to 500 mm of water. Out of many commercially available instruments a few have been described here. The main errors due to meniscus and capillary effects are minimised in all these instruments.

Prandtl-type Micromanometer

This U-tube manometer consists of a reservoir of large diameter and an inclined tube with two marks connected through a flexible tube. Two variants exist, in one (Fig. 6.4(a)) the reservoir is raised or lowered to restore the liquid level between the two marks, while in the other (Fig. 6.4(b)) the inclined tube is moved vertically. The capillary and meniscus errors are minimised by bringing the level to a reference null position (between two marks) before the application of the pressure. After the application of pressure either the reservoir or the inclined tube is moved vertically by a lead screw to achieve the null position again. The motion Δh of lead screw is used to calculate pressure difference or it may be directly indicated on the dial of the lead screw.

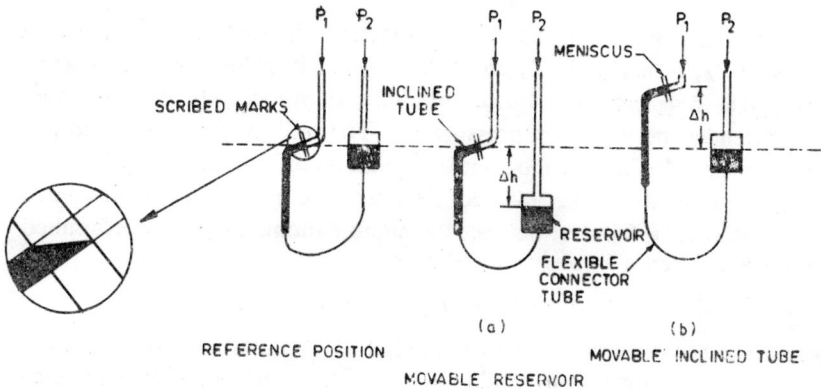

Fig. 6.4 Two variations of the Prandtl-type manometer

The pressure difference ($P_1 - P_2$) is obtained from the equation

$$P_1 - P_2 = W_M \Delta h$$

where W_M is the specific weight of manometer liquid. Figure 6.4 illustrates the operation of this micromanometer.

MICROMETER TYPE MANOMETER

In this type of micromanometer, meniscus and capillary effects are minimised by measuring liquid displacement with micrometer heads fitted with adjustable sharp index points located at or near the centre of large bore transparent tubes that are joined at their bases to form a U as shown in Fig. 6.5. The contact with the surface of manometer liquid may be sensed optically or electrically.

AIR MICROMANOMETER

An extremely sensitive, high response micromanometer uses air as its working fluid, and therefore avoids all capillary and meniscus effects

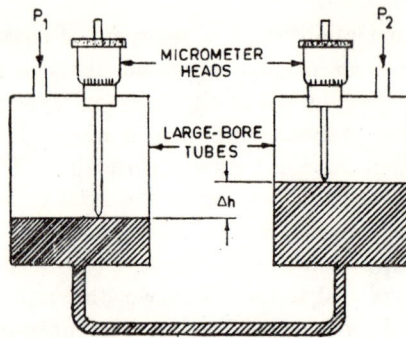

Fig. 6.5   A micrometer-type manometer

usually encountered in liquid manometry. In this device, reference pressure P_R, is mechanically amplified by centrifugal action in a rotating disc. The disc speed is adjusted until the amplified reference pressure just balances the unknown pressure P. The null position is obtained by observing the lack of movement of minute oil droplets sprayed into the glass indicator tube located between the unknown and amplified pressures (Fig. 6.6.). At balance the micromanometer yields the applied pressure difference through the relation

$$\Delta P = P_R - P = a\rho\omega^2$$

where ρ is the reference air density, ω is the angular speed of the disc, and a is a constant that depends on disc radius and annular clearance between the disc and housing. The measurement of pressure as small as 0.005 mm water can be made with this micromanometer with an uncertainty of 1%.

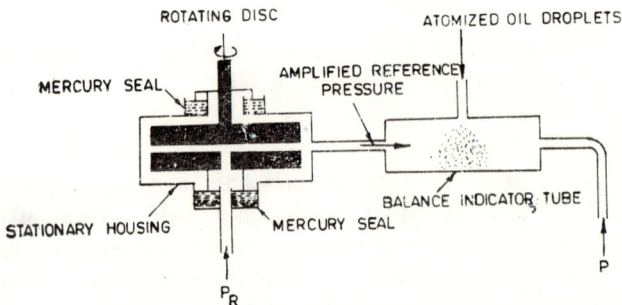

Fig. 6.6   An air-type centrifugal micromanometer

6.8   McLeod Gauge

It is a modified mercury manometer, used mainly for the measurement of vacuum pressures from 1 mm to 10^{-6} mm of Hg. It measures a differential pressure, and hence is very sensitive. Further, it measures absolute pressure because the pressure is given by the physical dimensions of the gauge. McLeod gauge is often employed for the calibration of electrical pressure gauges like Pirani and Penning gauges. The main limitations of this gauge are its slow response and extreme care required in its handling. The construction details and procedure of operation are given below:

A capillary C of a very uniform bore of cross-sectional area A is connected with a large bulb B. The vacuum pressure to be measured is connected as shown in Fig. 6.7. If the capillary contains vacuum, then as the reservoir is raised, the mercury level in tubes 1, 2 and 3 will rise and remain at the same level in all the three tubes till it reaches the end part of the tube 1. This is taken as the reference. The tubes 1 and 2 have the same bore dimensions to avoid surface tension effects.

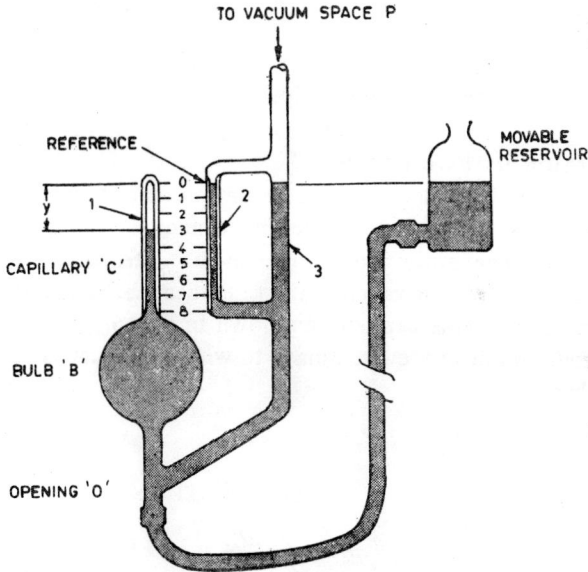

Fig. 6.7 The McLeod gauge (Non-linear scale)

The reservoir is then lowered till the mercury level is below O; the pressure source is thus connected to the capillary C. Therefore, the bulb and capillary are at the pressure of the source, which is to be measured. The reservoir is raised again, thus cutting off the pressure source from the bulb. The gas in the bulb is compressed and confined to the capillary as the reservoir is moved up. When the mercury level in tube 2 has reached the reference level, the height in the capillary is measured and pressure calculated using Boyle's law. Often the capillary readings are directly calibrated in pressure.

If the height in the capillary measured from the reference mark is y, the volume of the gas enclosed is Ay.

Let the volume of the bulb, capillary and the tube down to opening O be V_B, then

$$PV_B = AyP_C$$

where P_C is the pressure in the capillary and P is the unknown pressure. Further the pressure in the capillary 1 is

$$P_C = P + W_M y$$

Thus

$$P = \frac{AW_M}{(V_B - Ay)} \, y^2$$

Usually

$$Ay \ll V_B,$$

and hence

$$P = \frac{AW_M}{V_B} \, y^2$$

The pressure is thus obtained in terms of physical dimensions A and V_B. The tube 1 thus can be directly calibrated in pressure.

LINEARISATION OF PRESSURE SCALE

The scale, as indicated by this equation, is not linear but parabolic and hence the sensitivity of the instrument is not constant over the whole range. The scale, however, can be linearised. In this case, instead of raising the mercury to the reference mark, a constant volume of gas is trapped in the capillary as shown in Fig. 6.8. Let the volume of the capillary above the fixed mark to which mercury is always raised be V_C, then

$$PV_B = P_C V_C$$

$$= (P + W_M y) \, V_C$$

$$P = \frac{V_C W_M}{V_B - V_C} \, y$$

$$P \simeq \frac{V_C W_M}{V_B} \, y \; : \; (V_B \gg V_C)$$

Volumes V_C and V_B are known, hence pressure can be easily computed. Note that the pressure is linearly related to height y and so the scale is linear. In using a McLeod gauge it is important to realise that if the gas, whose pressure is monitored contains any vapour that are condensed by the compression process, pressure measurement becomes erroneous. Except for this effect the reading of McLeod gauge is not influenced by the composition of the gas. The main drawbacks of this gauge being lack of continuous output reading and limitations on the lowest measurable pressures. When it is used to calibrate other gauges, a liquid-air trap should be used between the McLeod gauge and tested gauge to prevent passage of mercury vapour.

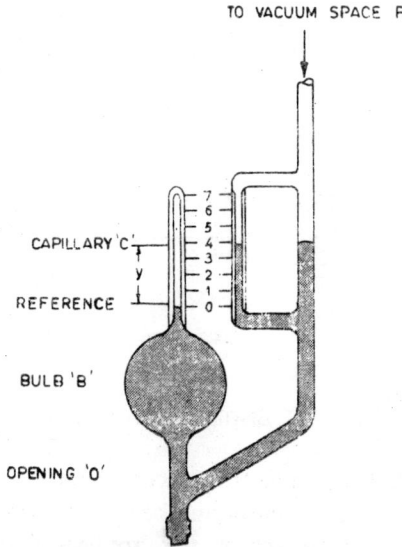

Fig. 6.8   The McLeod gauge (Linear scale)

A very compact design of a McLeod gauge is shown in Fig. 6.9.   Gas at low pressure is trapped in the manometer tube by mercury as the gauge rotates about a pivot by $\pi/2$ from horizontal position.   Mercury compresses the gas into the top of measuring tube where final volume, expressed as initial pressure is read off.

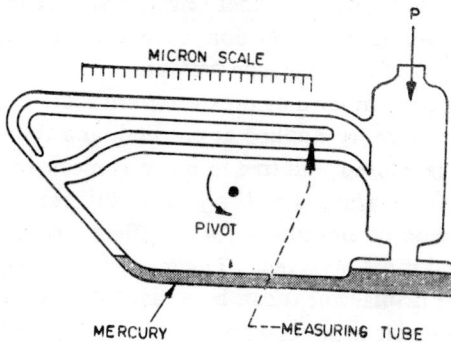

Fig. 6.9   The compact McLeod gauge

6.9   Pressure measurement with elastic transducers

Application of pressure to a body causes elastic deformations.   The magnitude of deformation can be related either analytically or experimentally to the applied pressure.   There are a wide variety of metallic elements that might be used as pressure transducers. Of these the most commonly used are some forms of Bourdon tubes, bellows or diaphragms. Some of these are shown in Fig. 6.10. Application of pressure causes a gross movement in these transducers. This movement may either directly actuate pointer/scale read out through suitable linkages or gears, or it may be transduced to an electrical signal.

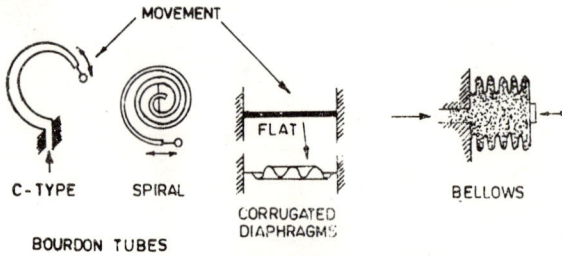

Fig. 6.10   Elastic pressure transducers

In the C-type Bourdon tube transducer, the elastic element is a small volume tube of oval cross-section and bent in C form. One end of the tube is fixed but open to accept the applied pressure while the other end is closed but free to allow displacement. Under pressure the oval cross-section of the tube tends to become circular consequently increasing the radius of curvature of the C-form. The movement can directly actuate the pointer through linkages or gears. The strain gauges can also be mounted to sense the displacement. There are other forms of Bourdon tubes having higher sensitivities. C-type Bourdon tube is used to measure pressures upto 10^7 mm Hg while the other versions are used upto 10^5 mm Hg due to higher sensitivity. The reference pressure in the case containing the Bourdon tube is usually atmospheric; the indicated pressure therefore is gauge pressure.

Another elastic element used as pressure transducer takes the form of bellows. The pressure is applied at one end, and the deflection of the other end which is closed but free to move is measured. The bellows can be paired to measure small pressure differences. The bellows transducer is generally unsuitable for transients because of the larger relative motion and mass involved. The movement of end can actuate the pointer, or the displacement can be measured by measuring changes in capacitance, inductance or resistance.

The diaphragms are widely used pressure transducers. They appear in the form of flat, corrugated or dished plates. The diaphragm transducers have a fairly good dynamic response. The central deflection of the diaphragm can be mechanically amplified and read off or can be sensed by strain gauges; capacitance and inductance measurements are also used for measuring the central displacement. Invariably mechanical amplification is used for corrugated diaphragms due to the relatively large central deflection. Because of the wider applications of diaphragm for pressure measurement, it is described here in detail.

The mathematical relation between pressure and central deflection for a flat circular diaphragm with its edge clamped is given by

$$P = \frac{16Et^4}{3a^4 (1 - \mu^2)} \left[\frac{y_c}{t} + 0.488 \left(\frac{y_c}{t} \right)^3 + \cdots \right]$$

where t, a and y_c are the thickness, radius and central deflection of the diaphragm respectively. E and μ are the Young's modulus and Poisson's ratio of the material of the diaphragm, and P is the applied pressure. The pressure deflection relation is therefore non-linear. In order to have a linear response the second and higher terms in the above expression must be smaller compared to the first term. If a non-linearity of 5% is acceptable, the maximum deflection must be less than 1/3 of the thickness. To facilitate linear response over a larger range of deflection that is imposed by the one-third thickness restriction, the diaphragm may be constructed out of corrugated discs.

MEASUREMENT OF CENTRAL DEFLECTION USING STRAIN GAUGES

Fig. 6.11(a) shows a cross-section of a diaphragm pressure pick up along with the arrangement of mounting the gauges. A flat circular diaphragm of radius 'a' and thickness 't' is rigidly clamped at the periphery, and resistance gauges R_1 and R_3 are mounted diametrically opposite at a radius of a_r while the resistances R_2 and R_4 are mounted very close to the center at a radius of a_t. The diaphragm is subjected to a uniform pressure P and the deflection is measured with the help of strain gauges. Both tension and compression stresses exist simultaneously and the mounting arrangement of gauges exploits this situation. The deflection-pressure relationship for this diaphragm is expressed by

$$y = \frac{3P (1 - \mu^2)}{16E\, t^3} (a^2 - r^2)^2.$$

This relation holds valid if the operation is within linear response region. The radial stress S_r and tangential stress S_t at any point on the low pressure side of the diaphragm when it is subjected to a uniform pressure P are given by

$$S_r = A \left[\left(1 + \frac{1}{\mu} \right) - \left(1 + \frac{3}{\mu} \right) \frac{r^2}{a^2} \right],$$

and

$$S_t = A \left[\left(1 + \frac{1}{\mu} \right) - \left(3 + \frac{1}{\mu} \right) \frac{r^2}{a^2} \right]$$

where A is a constant for a given pressure and is equal to

$$A = \frac{3Pa^2\mu}{8t^2}$$

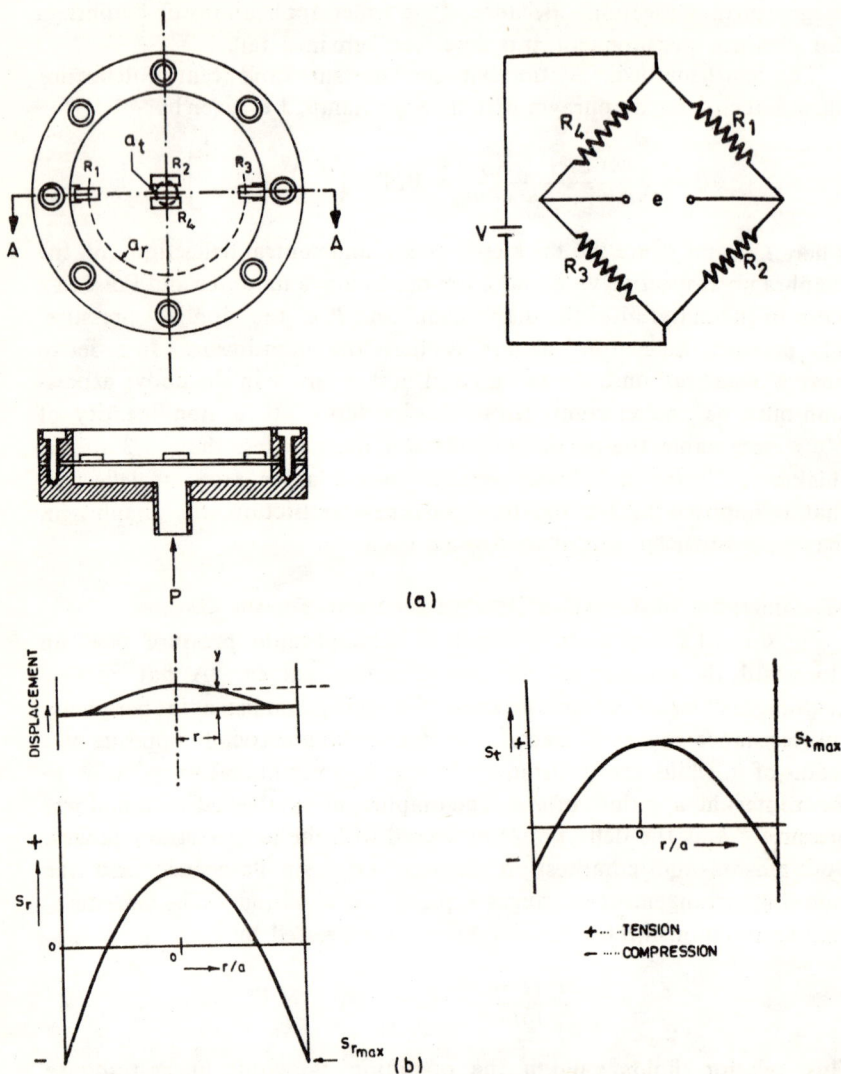

Fig. 6.11(a) Construction of a diaphragm transducer and mounting of strain gauges,
(b) Deflection and stress distribution over the diaphragm

The magnitude of tangential stress is maximum at the center ($r = 0$)
and has a positive value, i.e.

$$S_{t_{max}} = \frac{3Pa^2\mu}{8t^2}\left(1 + \frac{1}{\mu}\right),$$

while that of radial stress is maximum at the edge ($r = a$), and has a
negative value, i.e.

$$S_{r_{max}} = -\frac{3}{4}\frac{Pa^2}{t^2}.$$

The variation of S_t and S_r as a function of (r/a) is shown in Fig. 6.11 (b). The above equations are accurate only for sufficiently small pressures.

The scheme of mounting strain gauges which is very satisfactory and exploits the existence of both tensile and compressive stresses is illustrated in Fig. 6.11(a) To measure the stresses and at the same time provide temperature compensation, gauges R_2 and R_4 are placed as close to the centre as possible and oriented to read tangential strain since it is maximum at the centre, while the gauges R_1 and R_3 are mounted as close to the edge as possible and oriented to read radial strain as it is maximum at that periphery. It can be seen that the resistance changes occuring in all the four gauges due to the pressure give outputs which are additive while the resistance changes due to temperature give outputs that cancel out. The strain gauge arrangement, thus, provides temperature compensation.

The diaphragm is in a state of biaxial stress and both the radial and tangential stresses contribute to the radial and tangential strains at any point. The general biaxial stress-strain relations give

$$\epsilon_r = \frac{S_r - \mu S_t}{E}$$

and

$$\epsilon_t = \frac{S_t - \mu S_r}{E}.$$

The radial, S_r, and tangential, S_t, stresses can be calculated as a function of pressure from the parameters of the diaphragm. Using these values in the above equations, corresponding radial, ϵ_r, and tangential, ϵ_t, strains can be calculated. From the bridge sensitivity equations the output of the bridge is

$$e = \frac{V}{4}\left[\frac{\Delta R_1}{R_1} - \frac{\Delta R_2}{R_2} + \frac{\Delta R_3}{R_3} - \frac{\Delta R_4}{R_4}\right]$$

where e is the bridge output, V is the excitation voltage and R's and ΔR's are the gauge resistances and changes in the resistances due to application of pressure. Further ϵ_t and ϵ_r are measured respectively by gauges R_2, R_4 and R_1, R_3 which produce an additive output. Hence

$$e = \frac{VG_F}{4}(2\epsilon_r + 2\epsilon_t)$$

where G_F is the gauge factor. The output e can therefore be obtained from design data for a given pressure. Alternatively, the pressure transducer can be calibrated using known pressure source.

The dynamic response of the strain gauge bonded pressure transducers is good. A diaphragm can be excited into an infinite number of modes The frequency of the lowest mode is of interest for understanding its response to dynamic input. It can be shown that the frequency of the lowest mode of a edge-clamped diaphragm vibrating in vacuum (no fluid inertia effects) is given by

$$\omega_n = \frac{10.21}{a^2} \left(\frac{Et^2}{12\rho (1 - \mu^2)} \right)^{\frac{1}{2}} \text{ rad./sec};$$

where Young's modulus E is expressed in N/m^2, density ρ in Kg/m^3, and t and 'a' are in meters respectively. The natural frequency ω_n depends on the elastic properties of the material and the physical dimensions of the diaphragm. There are a number of factors such as liquid inertia, wrinkling of diaphragm, imperfect clamping etc., which may make the actual frequency value of the diaphragm different from that calculated by the above equation. As an example any wrinkling of the diaphragm sheet will increase its stiffness and hence raises its natural frequency.

Miniature pressure probes are now commercially available. A silicon diaphragm on which a Wheatstone bridge has been atomically bonded using diffusion techniques is the pressure sensing element. These are available in wide variety of ranges and possess very good frequency response.

6.10   Electrical methods

The pressure transducers which provide an output as an electrical signal fall under this group. A resistance pressure transducer which is used for the measurement of very high pressures is discussed first.

MEASUREMENT OF VERY HIGH PRESSURES—BRIDGMAN GAUGE

The high pressure range has been defined as beginning at 10^6 mm Hg and extends to 10^8 mm Hg. Very high pressures can be measured by resistance gauges. The pressure sensing element is a loosely wound coil of a relatively fine wire immersed in a pressure transmitting fluid. Application of pressure results in bulk compression effect that causes a change in the electrical resistance of the element. This change in resistance can be calibrated in terms of applied pressure.

The resistance of a wire is given by

$$R = \rho L/A = \rho L/CD^2$$

where ρ is the specific resistance of the material and C is a proportionality constant, its value being $\pi/4$ for circular cross-section; other symbols having their usual meaning.

On differentiation, one obtains

$$\frac{dR}{R} = \frac{d\rho}{\rho} + \frac{dL}{L} - \frac{2dD}{D}$$

The freely suspended wire in the pressure medium is subjected to biaxial stress condition only, because the ends, in providing electrical continuity, will generally not be subjected to pressure. Thus $S_x = S_y = P$

and $S_z = 0$, where S_x, S_y, S_z are stresses along x, y and z directions : z is taken along the length of the wire. Hence, the strain along x, y, z directions are given by

$$\epsilon_x = \epsilon_y = \frac{dD}{D} = -\frac{P}{E}(1 - \mu),$$

and

$$\epsilon_z = \frac{dL}{L} = \frac{2\mu P}{E}.$$

Substituting for dD/D and dL/L in the above equation, one gets,

$$\frac{dR}{R} = 2\frac{P}{E} + \frac{d\rho}{\rho}.$$

If the specific resistance ρ does not depend on pressure, $\dfrac{d\rho}{\rho}$ can be neglected and hence

$$\frac{dR}{R} = 2\frac{P}{E}.$$

Therefore

$$R = R_0(1 + bP),$$

where $b = 2/E$ is called the pressure coefficient of resistance. The resistance varies linearly with pressure.

The pressure transducer based on this principle is called Bridgman gauge. Figure 6.12 illustrates a typical gauge. The pressure sensing element is in the form of a loose coil enclosed in a flexible kerosene-filled bellows. One end of the coil is ground to the cell body and the other end, suitably insulated from the body, is brought outside for electrical connection. The pressure is transmitted to the sensing element through the kerosene-filled bellows. The resistance change brought about by the application of pressure is measured by some form of Wheatstone bridge.

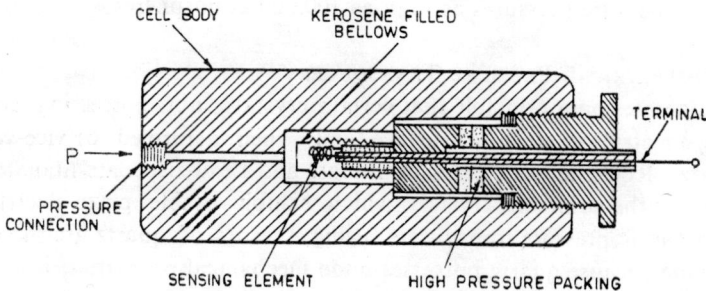

Fig. 6.12 The Bridgman gauge

Manganin and an alloy of gold are the two metals commonly used for pressure sensing element. Both metals provide linear output as a function of the applied pressure upto 3×10^4 Kgf/cm². Their pertinent characteristics are given in Table 6 3.

TABLE 6.3

Characteristics	Manganin 84 Cu, 12 Mn 4 Ni	Gold Chrome 97.9 Au, 2.1 Cr
Pressure sensitivity or pressure coefficient (ohm/ohm)/Kgf/cm^2	24.1×10^{-7}	9.55×10^{-7}
Temperature sensitivity (ohm/ohm)/°C	3.06×10^{-5}	1.44×10^{-6}
Resistivity ohm-cm	45×10^{-6}	2.4×10^{-6}

Although the pressure sensitivity of gold-chrome is lower than that of manganin, it is preferred in many cases because of its much smaller temperature sensitivity. Due to sudden pressure changes the kerosene used in bellows will experience a transient temperature change because of adiabatic compression. This will change the resistance of the sensing element in addition to that caused by pressure. This results in temperature error. It will be smaller for Gold-chrome sensing element.

Since the variation of resistance is associated with elastic movements within the wire, it occurs within the time required for sound wave to travel the wire radius. For a typical wire of 0.025 mm radius, this is of the order of 10^{-8} s. The dynamic response of the sensor is, therefore very good; the wire resistance changes with the application of pressure almost instantaneously. However, the accompanying temperature change will cause a transient error if temperature sensitivity is too high. Gauges of this type are commercially available to measure pressures upto 15000 Kgf/cm^2 with inaccuracy of 0.1 to 0.5%. They have been used to measure pressures as high as 100,000 atmospheres.

PIEZO-ELECTRICAL PRESSURE TRANSDUCER

It operates on a principle that certain crystals, not possessing center of symmetry, produce a surface emf when deformed or vice-versa. Quartz, Rochelle salt, Barium titanate and Lead-zirconate-titanate are some of the common crystals which exhibit usable piezo-electricity. Piezoelectric pressure transducers using sensors of quartz are the most common because quartz possesses good mechanical properties, is a good insulator, least influenced by moisture and has almost temperature independent response over a very wide range.

The pressure pick-ups made from these crystals are so designed that they show maximum piezo-electric response along the desired direction with no or very little response along the other direction. As an example, a quartz x-cut crystal of 2.5 mm thickness would have a sensitivity of about 10 volts/Kgf/cm². Such high sensitivity is typical of piezo-electric transducers. The use of piezo-electric transducer elements is primarily limited to the dynamic measurements. Hence, these elements are extensively used in sound pressure instrumentation, in acceleremeters and vibration pick-ups. Some commercially available systems using quartz transducers (very high leakage resistance) and electrometer input amplifiers (very high input impedances) achieve an effective total resistance of 10^{14} ohms which gives sufficiently slow leakage to allow static measurements.

Although the emf developed by piezo-electric transducers may be proportional to pressure, it is nonetheless difficult to calibrate them by normal static procedures. An attractive technique called 'electrocalibration' has been developed in which piezo-electric transducer is excited by an electric field rather than by an actual physical pressure to obtain calibration.

The dynamic response of piezo-electric pressure transducers is of second order. They possess very little internal damping ($\zeta \simeq 0.007$) and very high response frequency of the order of 100 kHz. They are essentially used for measuring rapidly fluctuating pressures of rapid transients.

6.11 Measurement of vacuum pressures

THERMAL CONDUCTIVITY GAUGE

At low pressures, kinetic theory of gases predicts a linear relationship between pressure and thermal conductivity. Conductivity of a gas is measured by measuring the temperature of the heated filament kept in the container filled with the gas. The temperature of the heated wire carrying the current will depend on two factors:

(i) the magnitude of current, and
(ii) the heat loss, both conductive and radiative.

The radiation losses can be minimised by using materials of low emissivity. The conduction loss depends on the composition of gas and hence calibration stays valid only for a particular composition.

The most common type of conductivity gauges are thermocouple, resistance (Pirani) and thermistor type and differ only in the way the temperature is measured. In thermocouple gauge, the temperature of the filament is measured by a thermocouple welded to it.

In resistance type gauges, the temperature is measured indirectly. As the temperature of wire changes, its resistance also changes. Thus, it is the resistance change which is a measure of temperature variation. The pressure is therefore measured in terms of resistance changes. The resistance change is measured by using a conventional Wheatstone circuitry. In order to minimise or compensate the effect of ambient temperature variations, a dummy gauge or a compensating cell is used as shown in Fig. 6.13. In thermistor type gauges, the resistance element is a thermistor—its resistance changes very rapidly with temperature. Because of its high sensitivity, compensation for ambient temperature variations is very important.

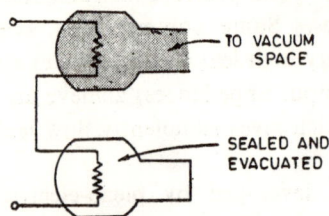

Fig. 6.13   Pirani gauge arrangements to compensate
for changes in ambient temperature

Ionisation Gauge

In an electric field, an electron can be accelerated such that it can ionise a gas molecule on collisions. The positively charged ions can be collected by a plate kept at a negative potential; an ion-current thus flows in the plate circuit. The pressure of the gas is proportional to the number of gas molecules and hence the ion current is also proportional to the pressure.

Figure 6.14 shows basic elements of an ionisation gauge. It is very similar to an ordinary triode electronic tube. It possesses a heated filament, a positively biased grid and a negatively biased plate, in an envelope connected to the pressure source. The electrons emitted from the filaments are accelerated by the grid voltage, they collide with gas molecules and ionise them. The positive ions are attracted to the plate, causing a flow of current i_i in the external circuit which is proportional to pressure. The electron current i_e which flows in the grid circuit is not effected by the secondary electrons due to ionisation and remains practically constant. The sensitivity of gauge is defined as

$$S = \frac{i_i}{i_e P}$$

where P is the pressure. Since i_e is practically constant, the sensitivity depends only on i_i. The input P and output i_i are therefore linearly related with in the region of operation of the gauge.

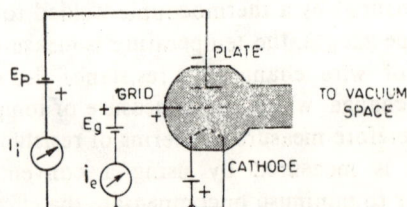

Fig. 6.14   Schematic of an ionisation gauge

As the pressure drops, ionisation current decreases as there are less number of molecules to be ionised. Ionisation current can be increased further if the traverse of the electrons from the filament is increased, thus increasing the probability of an electron colliding with a molecule. This is achieved by the application of a magnetic field perpendicular to the plane of paper so that the path of electrons is helical and hence the electrons meet more number of molecules in the path, causing higher ionisation current.

Ionisation gauges can be designed to measure vacuum pressure up to 10^{-12} torr. However, they suffer from the following two disadvantages:

(i) excessive pressure (1 to 2 μm) will cause rapid oxidation of filament and thus shorten its life, and

(ii) the electron bombardment is a function of filament temperature, thus requiring a careful control of the filament temperature.

These disadvantages are eliminated in 'alphatron' where a radium source is used to ionise the gas.

KNUDSEN GAUGE

This is another kind of gauge, that can be used to measure absolute pressure. Like McLeod gauge it shares the desirable feature of composition insensitivity but for the variation of accommodation coefficient from one gas to another. The accommodation coefficient is a measure of the extent to which rebounding molecule has attained the temperature of the heater surface. Further it is capable of giving continuous output readings.

Two vanes along with the mirror are suspended by a very fine filament; the restoring force is provided by the torsion in the filament (Fig. 6.15). Near the vanes are two heater plates, maintained at temperature T and are so arranged that one heater is in front of one vane and the other behind the second vane. The separation between the plate and vane is less than the mean free path of the surrounding gas, which is at temperature T_g. The vanes are at the gas temperature.

The molecules striking from the heater side impart a higher momentum due to being at higher temperature than from the other side. Thus there is a net momentum imparted to the vanes, causing them to rotate about the suspension. The rotation is monitored by a light pointer. The total momentum change depends on molecular density, which in turn is related to the pressure and the temperature of the gas. An expression for the pressure may be derived in terms of the measured force F and temperature T and T_g as follows:

The velocity of molecules at T_g is $v_0 = \sqrt{3kT_g/m}$, and the velocity of molecules at T is $v_1 = \sqrt{3kT/m}$, where m is the mass of a single molecule and k is the Boltzmann constant. Thus

$$m(v_1 - v_0) = \sqrt{3km}(\sqrt{T} - \sqrt{T_g}).$$

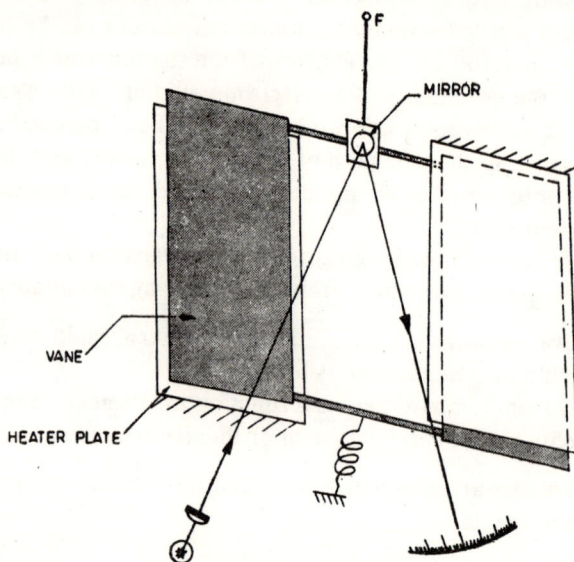

Fig. 6.15 Knudsen gauge

Let the molecular density be n, therefore $n/6$ molecules are moving in one directions and their velocity is v_0. Thus, the number of molecules which hit the vane per second is $(n/6)\, v_0$. Therefore, the rate of momentum transfer to the vane is

$$\left(\frac{n}{6}\right) v_0 \sqrt{3km} \left(\sqrt{T} - \sqrt{T_g}\right) = F.$$

on substitution for v_0, one has

$$F = \frac{nkT_g}{2} \left(\sqrt{\frac{T}{T_g}} - 1\right)$$

Writing $P = nkT_g$,

$$F = \frac{P}{2} \left(\sqrt{\frac{T}{T_g}} - 1\right)$$

For smaller temperature difference $(T - T_g) \ll T$,

$$\sqrt{\frac{T}{T_g}} = \sqrt{1 + \frac{T - T_g}{T_g}} = 1 + \frac{T - T_g}{2T_g}.$$

Therefore

$$F = \frac{P}{4} \frac{T - T_g}{T_g} = K\phi$$

where ϕ is the rotation and K is a torsional constant. The pressure is thus known in terms of measurable quantities.

The Knudsen gauge is a suitable device for calibrating other pressure gauges between 10^{-2} to 10^{-8} torr.

6.12 Dynamics of pressure transducers

When the pressure transducer is directly exposed to the fluid pressure to be measured, as in certain piezo-electric transducers, the system's dynamic characteristics are those of the transducer itself. However, the majority of pressure measurements involve fluid transmission of the pressure signal through various tubes and chambers from the point of interest to pressure transducer. For static pressure measurement there is no influence of pressure transmitting elements on the measurement. However, for fluctuating pressure signals the dynamic response of pressure measuring instrument is governed by the following two factors:

(i) response of the transducer, and
(ii) response of the pressure transmitting fluid and tubing etc.

The frequency response of the pressure-transmitting fluid and tubing etc. often determines the overall response. Further the response depends on diameter, length of tubings, volume of pressure chambers, pressure differences, etc.

To illustrate the dynamic response of a pressure system with a very simple analysis which is restricted to small pressure differences, steady state laminar flow conditions are assumed to exist. Consider the pressure transmission through a tube of length L and diameter $2r$. This is connected to a chamber of volume V; pressure transducer is assumed to be located in the chamber. Let P_i be the input pressure and P_m be the measured pressure. P_i is assumed to vary in some fashion and it is desired to know how P_m changes. Initially $P_i = P_m = P_0$. However, from now on P_i and P_m refer to pressures over and above P_0. Due to a pressure signal P_i a slug of gas moves in the system. The following forces are acting in the system:

(i) A force of magnitude $\pi r^2 P_i$ due to pressure P_i,
(ii) The viscous force due to wall shearing stress of magnitude $8\pi\mu L\dot{x}$, where x is the displacement of the slug of gas in the tube, and μ is the fluid viscosity.
(iii) If the slug of gas moves by x into the volume V of the chamber, the pressure P_m will increase. Assuming that compression occurs under adiabatic conditions, the adiabatic bulk modulus E_a of gas is given by

$$E_a = -\frac{dP}{(dV/V)} = \gamma P$$

where γ is the ratio of specific heats. The volume change dV due to the motion of slug is $dV = \pi r^2 x$. Therefore, the excess pressure P_m is $(\pi r^2 E_a/V)$. The force due to this pressure is $(\pi^2 r^4 E_a/V)$.

(iv)   The force due to acceleration of gas is given by

$$\frac{4}{3} \pi r^2 \rho L \ddot{x},$$

where ρ is the density of the gas.   The effective mass of the gas is taken as (4/3) of the actual mass.

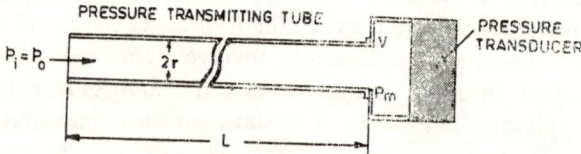

Fig. 6.16   Schematic of pressure transmitting system

Therefore the force balance equation is written as

$$\pi r^2 P_i - 8\pi\mu L\dot{x} - (\pi^2 r^4 E_a / V) = \frac{4}{3} \pi r^2 \rho L \ddot{x}.$$

Since $P_m = (\pi r^2 E_a / V)\, x$, the above equation is rewritten as

$$\frac{4}{3} \frac{\rho}{E_a} \frac{VL}{\pi r^2} \ddot{P}_m + \frac{8\mu LV}{\pi r^4 E_a} \dot{P}_m + P_m = P_i.$$

This is a second order equation.   If an input pressure wave of frequency ω is impressed over this system, its frequency response will be given by

$$\left|\frac{P_m}{P_i}\right| = \frac{1}{\{[1 - (\omega/\omega_n)^2]^2 + 4\zeta^2\,(\omega/\omega_n)^2\}^{1/2}}$$

where the natural frequency ω_n and damping coefficient ζ are defined as

$$\omega_n = \frac{r}{2}\sqrt{\frac{3\pi E_a}{VL\rho}},$$

and

$$\zeta = \frac{2\mu}{r^3}\sqrt{\frac{3LV}{\pi E_a \rho}}.$$

The phase angle for the pressure signal is

$$\phi = \tan^{-1}\frac{-2\zeta\,(\omega/\omega_n)}{1 - (\omega/\omega_n)^2}.$$

When the chamber volume is relatively large, and the transmission tube is of very small diameter, the system's characteristics will approach those of first order system i.e.

$$\left|\frac{P_m}{P_i}\right| = \frac{1}{[1 + 4\zeta^2\,(\omega/\omega_n)^2]^{1/2}}$$

In this case the chamber is analogous to a capacitor, and the tube analogous to a resistor. Such a system called an acoustical filter can be designed to attenuate frequencies above any given value.

6.13 Considerations for pressure gauge calibration

STATIC CALIBRATION

The familiar dead weight tester may be used to provide reference pressures with which transducer outputs may be compared. The testers of this type are useful upto 10^5 mm Hg, and by the use of special designs the limit may be extended to 10^6 mm Hg. The linearity of Bridgman gauge may be used to extend the range beyond this. For medium pressures, the precision mercury column (manometers) are used as calibration standards.

For vacuum pressures in the range of 10^{-1} to 10^{-3} torr McLeod gauge is considered as standard. A pressure dividing technique is used for measuring pressures below 10^{-3} mm Hg. The technique uses flow through a succession of accurate orifices: the low down stream pressure is related to the higher up stream pressure which is measured by McLeod gauge. The technique can be further improved by substituting ionisation guage for McLeod guage. The ionisation guage is calibrated at least at one point say 0.1 torr and its known linearity is used to extend the range down to 10^{-7} torr.

DYNAMIC CALIBRATION

The dynamic response characteristics of a pressure transducer are obtained either by impulse, step or frequency response tests; the step function tests being perhaps the most common.

Step function tests are used for systems which are not used at frequencies greater than 1 kHz. The step input is obtained by bursting a thin diaphragm subjected to a high gas pressure. A general rule for step testing is that the rise time of the step function must be less than one-fourth of the natural period of the system to be tested if it is to excite natural oscillations. For pick-ups with frequency greater than 1 kHz, shock tube is used to provide step input.

Exercises

1. Show that the dynamic response of an idealised manometer is described by a second order equation. Express its natural frequency and damping coefficient in terms of the parameters of the manometer.
2. A well type manometer is to be used to measure differential pressure in a water flow system. The manometer uses a special bromide liquid of specific gravity of 2.95. The well and the tube are of 75 mm and 5 mm diameters respectively. The scale placed along the tube has no correction factor for the area ratio of the manometer. Calculate the value of this factor which must be multiplied by the manometer reading in mm to find the pressure differential in Kgf/cm^2.
3. A pressure pick-up has the following specifications:

$a = 80$ mm, $E = 1.12 \times 10^{11}$N/m², gauge resistance $= 120\ \Omega$

$a_r = 60$ mm, $\mu = 0.3$, gauge factor $= 2.0$

$a_t = 10$ mm, battery voltage $= 5.0$ V.

$t = 1$ mm, $\rho = 7.93$ gm/cm³

(a) Calculate the sensitivity in mV/ Kgf/cm².
(b) What is the natural frequency of the lowest mode in vacuum?
(c) What is the maximum allowable pressure for 2% non-linearity? What is the voltage output at this pressure?

4. Design a bonded strain gauge pressure transducer (flat diaphragm) and the bridge circuit to meet the following requirements:

Maximum pressure $= 10$ Kgf/cm²

Natural frequency in vacuum $= 10$ Hz minimum

Maximum non-linearity $= 3\%$

Full scale output $= 10$ mV

The diaphragm is to be made from stainless steel ($E = 2 \times 10^{11}$ N/m², $\mu = 0.26$). The strain gauges of 300Ω resistance, gauge factor of 2.0 and size of 8 mm by 8 mm are to be used.

5. A bonded strain gauge diaphragm of 60 mm diameter is to be constructed of phosphor bronze ($E = 1.12 \times 10^{11}$ N/m², $\mu = 0.3$) to measure a maximum pressure of 15 Kgf/cm². Calculate the thickness of the diaphragm so that the maximum deflection is one-third of the thickness. Calculate its natural frequency in vacuum.

6. Calculate the resistance change of 100Ω coil of both manganin and gold-chrome materials respectively for 3.5×10^8 N/m² pressure and 40°C temperature changes.

7. A bridgman gauge has an active element in the form of a manganin wire of 25 µm diameter and 25 mm long housed in a channel 12 mm in diameter by 10 mm deep as shown in Fig. 6.17. It is to be used to measure short duration pressure pulses of strength in the range of 3.5×10^8 N/m² to 7.0×10^8 N/m². The pressure pulse will be accompanied by a temperature pulse of the order of 500°C.

(a) Design a Wheatstone bridge circuit for this unit and estimate the system sensitivity.
(b) Estimate the frequency response with (i) the system as shown, and (ii) the cavity filled with silicon grease.

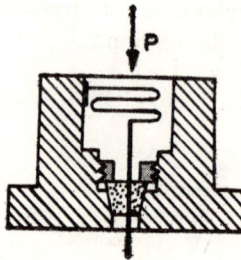

Fig. 6.17

8. A McLeod gauge has a capillary of 1 mm diameter and a bulb of 100 c.c. Calculate the pressure indicated by a reading of 20 mm. What error would result in the measurement if the volume of capillary is dropped in comparison with the volume of the bulb?

9. A Knudsen gauge is to be designed to operate at a maximum pressure of 1.0 μm. For this application the spacing of the vane and the plate is to be less than 0.3 mean free path at this pressure. Calculate the force on the vanes at pressures of 1.2 μm and 0.02 μm, when the gas temperature is 293°K and temperature difference is 50°K.

10. A small tube of 8 cm length and 0.5 cm diameter is connected to a pressure transducer through a volume of 3 cm³. Air at NTP is the pressure transmitting fluid. Calculate the natural frequency and damping ratio of the system. What would be the attenuation of 40 Hz pressure signal in the system?

Flow Measurement and Flow Visualisation

7.1 Introduction

The measurement of fluid is not only necessary in the research laboratories but is used in almost all types of industries, and encampasses a varied applications ranging from measurement of blood flow rate in the human artery to the measurement of the flow of liquid oxygen in a rocket. The degree of difficulty in measurement of flow is furthered by environmental conditions and requirements of measurement. In most of the cases the flow measurement requires the measurement of both the pressure (or pressure differential) and temperature. The physical and chemical properties of the fluid make the measurements more difficult. But the two phase flow, with dramatic changes in volume and energy, probably presents the greatest problem. However the two component flow (gas/liquid and slurries) is not so difficult to meter, although the distribution of mass and kinetic energy within the flow is non-uniform. The problems associated with the flow measurement are so varied and complex that every one concerned with its measurement must know the requirements in advance. No attempt has been made here to discuss these requirements with reference to a particular flow measurement situation.

The basic principles of flow measurement of most commonly used instruments are discussed in this chapter. Flow visualisation techniques that offer full field picture of the flow are also presented.

7.2 Types of flow-measuring instruments

Instruments used in the measurement of flow may be categorised into two main classes:

Quantity Meters: In this class of instruments, total quantity which flows in a given time is measured and an average flow rate is obtained by dividing the total quantity by time.

Flow Meters: In this class of instruments, actual flow rate is measured. Flow rate measurement devices frequently require accurate pressure and temperature measurements in order to calculate the output of the instrument. The overall accuracy of the instrument depends on the accuracy of pressure and temperature measurements. Quantity meters are used for calibration of the flow meters.

The classification can be further done as follows.

Quantity Meters
  (a) Weight or volume tanks
  (b) Positive displacement or semi-positive displacement meters.
Flow Meters
  (a) *Obstruction meters*
      (i) Orifice
      (ii) Nozzle
      (iii) Venturi
      (iv) Variable-area meters
  (b) *Velocity probes*
      (i) Static pressure probes
      (ii) Total pressure probes
  (c) *Special methods*
      (i) Turbine type meters
      (ii) Magnetic flow meters
      (iii) Sonic flow meters
      (iv) Hot wire/film anemometers
      (v) Mass flow meters
      (vi) Vortex shedding phenomenon
  (d) *Flow visualisation methods*
      (i) Shadowgraphy
      (ii) Schlieren photography
      (iii) Interferometry.

This does not exhaust the list of flow-measuring systems but does attempt to include those systems which are of general interest. Special emphasis will be given to those instruments which find importance both in research and industry. The range, accuracy, operating conditions and other parameters of some of these instruments are given in Table 7.1.

7.3 Quantity meters

These types of meters give an indication which is proportional to the total quantity that has flown in a given time. They are used for the flow measurement of both liquids and gases. A wide variety of these instruments are available but only positive displacement type instruments are discussed here. They all 'chop' the flow into 'pieces' of known size (known volume) and then count the number of 'pieces'. Some of the configurations currently in use include reciprocating piston, reciprocating diaphragm, helical impellers, revolving vane, rotating drum, rotating disc and lobed impellers. Figure 7.1 shows a few of the positive displacement meters.   These are often used for the measurement of volumetric flow of water, high viscosity liquid and liquids of varying viscosity. Their accuracy can be very good even at low end of the flow range. Since they are devices with moving parts, their accuracy may suffer with time due to wear.   The accuracy can be enhanced by machining moving parts with fine clearances.   Resistance to corrosive liquids can be increased by using special materials, both for the case and moving parts.   Measurement of liquid with entrained vapour can be a very big problem, which is solved by inserting good vapour traps in the flow pipe.   Most of these instruments are totalizers and do not attempt to measure instantaneous flow rates.   The flow is transduced to rotary/ linear motion.

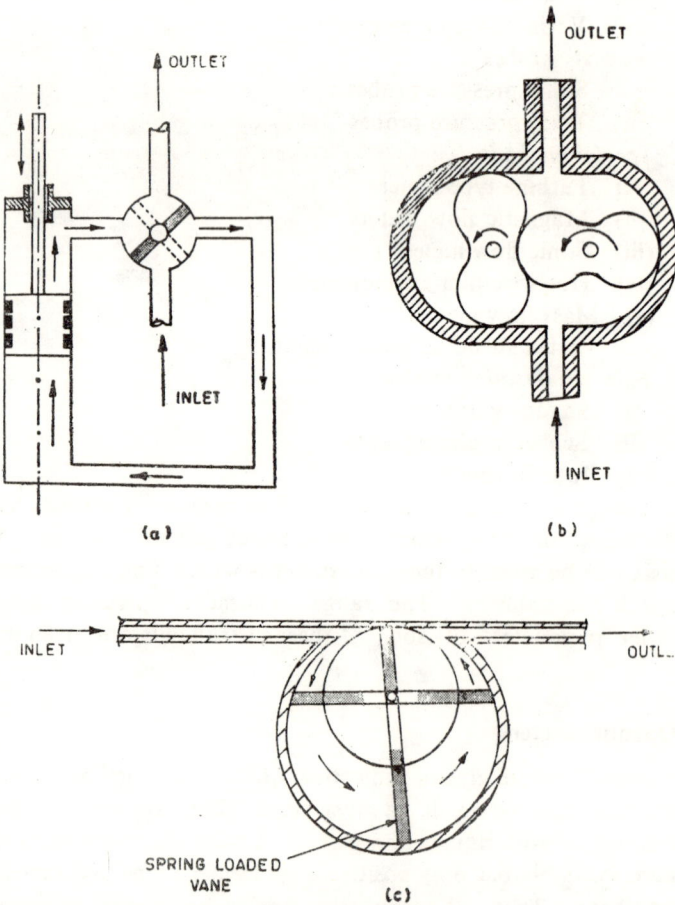

Fig. 7.1 Schematic of
(a) Reciprocating Piston
(b) Lobed Impellers and
(c) Rotary Vane Flowmeters.

In short one can say that displacement meters are hydraulic or pneumatic motors whose cycles of motion are recorded by some form of counter. Energy is extracted from the flow to drive these meters, resulting in a pressure loss from inlet to exit of the instruments. But the energy required is extremely low, just enough to overcome friction in the system.

7.4 Flow meters

OBSTRUCTION METERS OR HEAD METERS

When a fluid flows through a pipe with a restriction, the velocity of flow increases due to the decrease in area. When a fluid flow passes an obstruction to it, it suffers a loss in its static pressure. This loss in static pressure has a direct relation with the velocity of flow thereby the principle of conservation of energy. The obstruction may be a restriction, a bend or of any other form. The pressure drop is an indication of the flow rate. Thus the obstruction tranduces the velocity into a pressure change. Consider a one-dimmensional flow in a pipe as shown in Fig. 7.2. The continuity equation for this situation demands that

$$\dot{m} = \rho_1 A_1 u_1 = \rho_2 A_2 u_2$$

where m is mass flow rate, ρ's the densities at planes 1 and 2 and u's the flow velocities. If the flow is adiabatic and frictionless and fluid is incompressible ($\rho_1 = \rho_2 = \rho$), the flow is governed by the Bernoulli equation which may be expressed as

$$p_1 - p_2 = \frac{\rho}{2g_c} (u_2^2 - u_1^2),$$

where p_1 and p_2 are the pressures at planes 1 and 2 respectively. Substituting for u_1 in the above equation, one obtains

$$p_1 - p_2 = \frac{\rho u_2^2}{2g_c} \left(1 - \left(\frac{A_2}{A_1} \right)^2 \right)$$

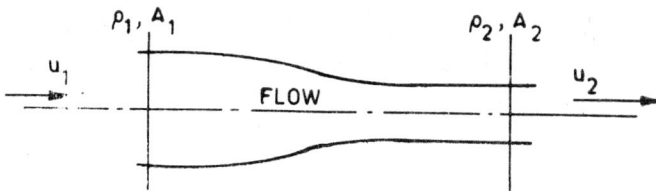

Fig. 7.2   General one dimensional flow system

The volumetric flow rate is given by

$$\dot{Q} = A_2 u_2 = \frac{A_2}{[1 - (A_2/A_1)^2]^{1/2}} \sqrt{\frac{2g_c}{\rho} (p_1 - p_2)}$$

The volumetric flow rate is thus proportional to the square root of pressure drop and is a function of other known parameters. Therefore, a channel like the one shown in Fig. 7.2 can be used for flow measurement by measuring pressure differential.

Discharge Coefficient

In the derivation of the above equation, it is assumed that the channel is frictionless. However, no channel is frictionless and some losses are always present in the flow. The actual flow rate is then different from that calculated on the basis of the above equation. The actual flow rate is related to the ideal one through the following relation:

$$C = \frac{\dot{Q}_{\text{actual}}}{\dot{Q}_{\text{ideal}}}$$

where C is known as discharge coefficient. The discharge coefficient is not a constant and may depend strongly on Reynolds number ($< 15{,}000$) and the channel geometry.

Compressible Fluid

The flow of a compressible fluid, say an ideal gas, obeys the following equation of state

$$p = \rho RT$$

where R is the gas constant and T is the absolute temperature. For reversible adiabatic flow, the steady flow energy equation is

$$c_p T_1 + \frac{u_1^2}{2g_c} = c_p T_2 + \frac{u_2^2}{2g_c}$$

where c_p is the specific heat at constant pressure and is a constant for an ideal gas. Assuming a very small approach velocity ($u_2^2 > u_1^2$) and applying the continuity and state equations to the above equation, the mass flow rate $\dot{m}$ is given by

$$\dot{m}^2 = 2g_c A_2^2 \frac{\gamma}{\gamma - 1} \cdot \frac{p_1^2}{RT_1} \left(\frac{\rho_2^2}{\rho_1^2} - \frac{T_2}{T_1} \frac{\rho_2^2}{\rho_1^2} \right)$$

where $\gamma = c_p/c_v$ is adiabatic constant. This equation can be rewritten as

$$\dot{m}^2 = 2g_c A_2^2 \frac{\gamma}{\gamma - 1} \frac{p_1^2}{RT_1} \left[\left(\frac{p_2}{p_1} \right)^{2/\gamma} - \left(\frac{p_2}{p_1} \right)^{\frac{\gamma + 1}{\gamma}} \right]$$

Compressible vs. Incompressible Fluid Flow

The above equation can be simplified if the pressure drop Δp ($= p_1 - p_2$) is less than $p_1/4$. In this case, it is given by

$$\dot{m} = \sqrt{\frac{2g_c}{RT_1}} A_2 \left[p_3 \Delta p - \left(\frac{1 \cdot 5}{\gamma} - 1 \right) \Delta p^2 + \cdots \right]^{1/2}$$

However, if the pressure drop is very small such that $\Delta p < p_1/10$, a further simplification results as the second term can be neglected in comparison with the first term. Therefore

$$\dot{m} = \sqrt{\frac{2g_c}{RT_1}} \, A_2 \, \sqrt{p_2 \, \Delta p}.$$

Comparing this with the expression for incompressible fluid flow, it may be concluded that for small values of pressure drop, Δp, compared with p_1, the flow of a compressible fluid may be approximated by the flow of an incompressible fluid.

TYPES OF RESTRICTION METERS

The restriction provided in the flow passage for the purpose of flow metering is the primary element in the flow meters. Several types of these exist but the three popular types are called, Orifice plate, Flow nozzle, and Venturi tube. These are shown in Fig. 7.3 along with the curves for pressure variation along the channel. The flow meters based on these are also called head meters.

Orifice Plate

It is a thin, flat disc, with an orifice for the passage of fluid, and is inserted between the flanges in the pipe. It can be readily rebored or replaced to accommodate flow capacity changes. Generally the orifice is concentric; it is sometimes provided with an additional small hole for the passage of condensates and gases. When gas is metered, the hole is located at the bottom to allow the condensates to pass in order to prevent its build-up at the orifice. When the liquid is metered, the extra hole is at the top to permit the gas to pass, thus avoiding the build-up of gas pockets. The hole may introduce measurement errors. Therefore an orifice in eccentric or segmental form is used. When metered liquids contain high percentage of dissolved gases or metered gases carry condensates, the eccentric type orifice plate is recommended.

The orifice may be installed at the bottom of a horizontal or sloping pipe line to permit the free passage of condensates, or at the top of the pipe line to permit the free passage of gases carried in the flowing fluid. The orifice plates of moderate size are least expensive. The main disadvantage, however, of this kind of restriction is a permanent pressure loss, often upto 30%.

The flow rate calculations for the orifice plate are made on the basis of the following equations:

Incompressible flow

$$\dot{Q}_{actual} = KA_2 \left(\frac{2g_c}{\rho} (p_1 - p_2) \right)^{1/2}$$

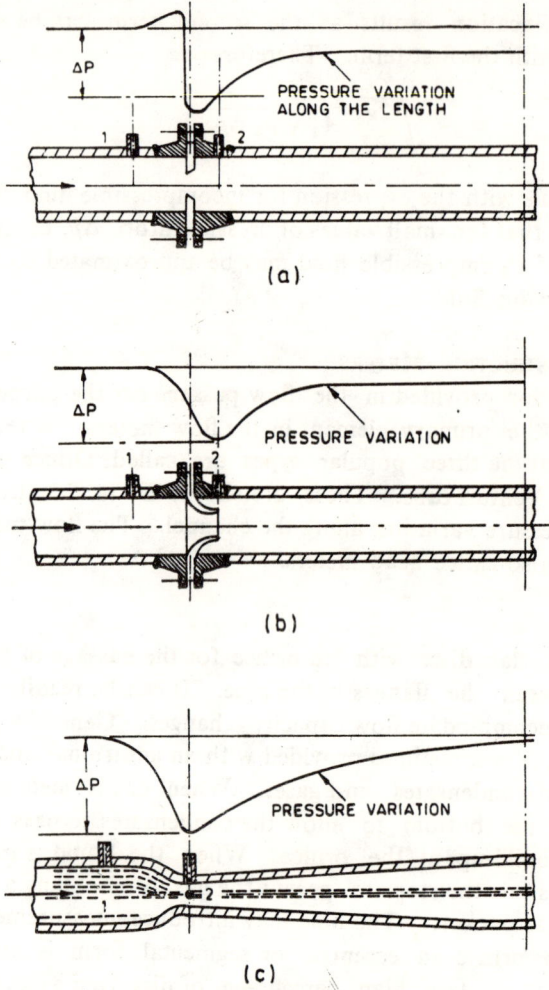

Fig. 7.3   (a)   An orifice plate
(b)   A flow nozzle
(c)   A venturi

Compressible Flow

$$\dot{m}_{\text{actual}} = YKA_2 \left(2g_c\rho_1 \, (p_1 - p_2) \right)^{1/2}$$

where the flow coefficient K and expansion factor Y are defined as

$$K = CM = C \frac{1}{[1 - (A_2/A_1)^2]^{1/2}}$$

and

$$Y = Y_1 = 1 - [0.41 + 0.35 \, (A_2/A_1)^2] \left(\frac{p_1 - p_2}{\gamma p_1} \right)$$

when either the flange taps or vena contracta taps are used, and

$$Y = Y_2 = 1 - (0.333 + 1.145\,(\beta^2 + 0.7\beta^5 + 12\beta^{13})]\left(\frac{p_1 - p_2}{\gamma p_1}\right)$$

when pipe taps are used. The constants M and β are the velocity of approach factor and diameter ratio $\sqrt{A_2/A_1}$, respectively.

Flow Nozzle

The flow nozzle is supported between standard flanges. The rounded approach has a curvature equivalent to the quadrant of an ellipse. The curved surface guards the nozzle from corrosive/erosive effects due to the suspensions in the gas, thus contributing to its long life. It allows measurement of flow rates which are about 60 to 65% higher than the maximum flow rate for which an orifice plate can be used. The flow nozzle is mainly used for metering fluids flowing under high pressures through lines of minimum size due to some reason. Another advantage of using nozzle is that it requires smaller straight piping before and after the primary element compared to that of the orifice. The permanent pressure loss has a magnitude between that of an orifice and a venturi. Its existence is attributed to the absence of recovery cone.

The flow rate calculations are made on the basis of following equations :

Incompressible Fluids

$$\dot{Q}_{actual} = KA_2\sqrt{\frac{2g_c}{\rho}(p_1 - p_2)}$$

Compressible Fluids

$$\dot{m}_{actual} = YKA_2\sqrt{2g_c\rho_1(p_1 - p_2)}$$

where the expansion factor Y is given by

$$Y = \left[\left(\frac{p_2}{p_1}\right)^{2/\gamma}\left(\frac{\gamma}{\gamma - 1}\right)\left(\frac{1 - (p_2/p_1)^{(\gamma-1)/\gamma}}{1 - (p_2/p_1)}\right)\frac{1 - (A_2/A_1)^2}{1 - (A_2/A_1)^2\,(p_2/p_1)^{2/\gamma}}\right]^{1/2}$$

Venturi Tube

The venturi tube offers the best accuracy, least pressure loss ($\simeq 13\%$) and best resistance to abrasion and wear from dirty fluids. It is however, expensive and occupies substantial space. Due to its excellent pressure recovery characteristics its use is recommended where measuring conditions require extremely low pressure loss. Because of its streamlined approach and exit, use of venturi tube is often considered when the flow of liquid with solids in suspension must be metered. The pressure differential that is being measured exhibits sensitivity to the concentration of the suspension in the liquid. There is an appreciable distance between the pressure taps in a venturi tube; the fluid friction along a side wall of pipe which varies as the fifth power of the diameter can have significant effect on the pressure differential. Therefore, when suspensions are carried in the fluid, the fluid friction depends on the concentration of suspension and hence pressure differential will exhibit sensitivity to the concentration of suspensions.

The flow rate calculations for venturi tube are made on the basis of following equations:

Incompressible Fluids

$$\dot{Q}_{actual} = KA_2 \sqrt{\frac{2g_c}{\rho}(p_1 - p_2)}, \text{ and}$$

Compressible Fluids

$$\dot{m}_{actual} = KYA_2\sqrt{2g_c\rho_1(p_1 - p_2)}, \text{ and}$$

where the expansion factor Y is given by the equation used for the flow nozzle.

PRACTICAL CONSIDERATIONS

The construction of obstruction meters has been standardised by organisations like ASME, DIN and ISI. The recommendations are briefly discussed here:

Orifice

The recommended installations for concentric, thin plate, square edged orifice are shown in Fig. 7.4 (a). Note three standard pressure-tap locations are used:

(i) *Flange Taps:* The taps are made in the flanges. This installation is universally used.

(ii) *Pipe Taps:* The inlet pressure tap is located one pipe diameter upstream and the outlet tap is located half diameter down stream of the orifice.

(iii) *Vena Contracta Taps:* The upstream pressure tap is located one pipe diameter upstream of the orifice plate, and the downstream pressure tap is located at vena contracta. Such taps are employed mostly in large size pipes where the use of a flange union is impractical.

The orifice discharge coefficient is sensitive to the condition of the upstream edge of the hole. The discharge coefficient is the same for liquids or gases as long as the Reynolds number is same. Figure 7.5 shows the dependence of flow coefficient, CM, on Reynolds number for various values of β, the diameter ratio, for flat-plate orifice.

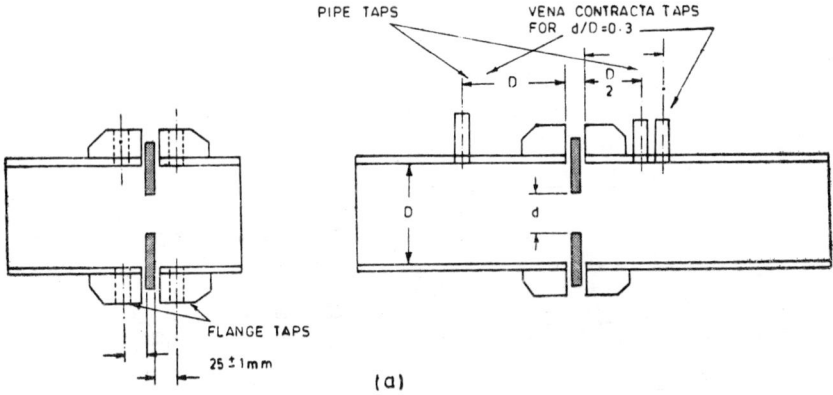

(a)

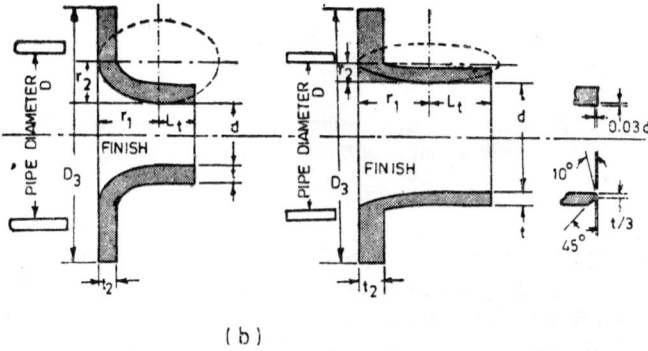

(b)

Low β Series: $0.20 \leqslant \beta \leqslant 0.5$, High β Series: $0.25 \leqslant \beta \leqslant 0.8$, Optimum Designs

$$r_1 = d, \ r_2 = 2d/3, \ L_t = 0.6 \, d, \ r_1 = D/2, \ r_2 = \left(\frac{D-d}{2}\right), \ L_t = 0.6d$$

$3 \text{ mm} \leqslant t \leqslant 13 \text{ mm}$	$3 \text{ mm} \leqslant t \leqslant 18 \text{ mm}$
$3 \text{ mm} \leqslant t_2 \leqslant 0.15D$	$3 \text{ mm} \leqslant t_2 \leqslant 0.15 \, D$

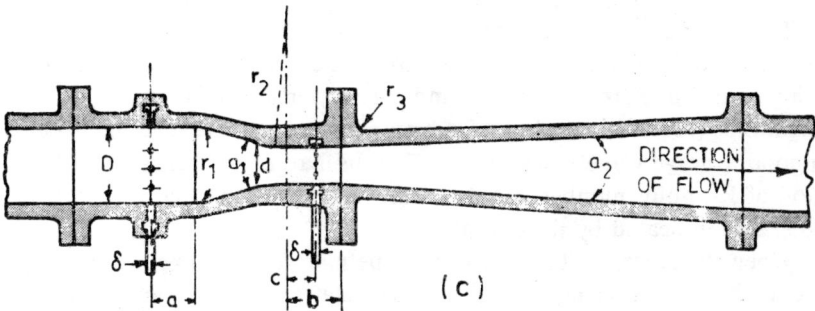

(c)

D = Pipe Diameter inlet and outlet, d = throat dia as required
a = 0.25D to 0.75D for 100 mm $\leqslant D \leqslant$ 150 mm, 0.25D to 0.50D for
150d mm $\leqslant D \leqslant$ 800 mm
$b = d, \ c = d/2, \ r_1 = 0$ to $1.375D, \ r_2 = 3.5$ to $3.75d., \ r_3 = 0$ to $0.25d$
δ = 6 mm to 12 mm depending on D. Annular Pressure Chamber with 4
Piezometer vents
$\alpha_1 = 21° \pm 1°, \ \alpha_2 = 7°$ to $15°$

Fig. 7.4 Recommended installations: (a) Concentric thin plate square
edged orifice, (b) Long radius flow nozzles, (c) Venturi tubes.

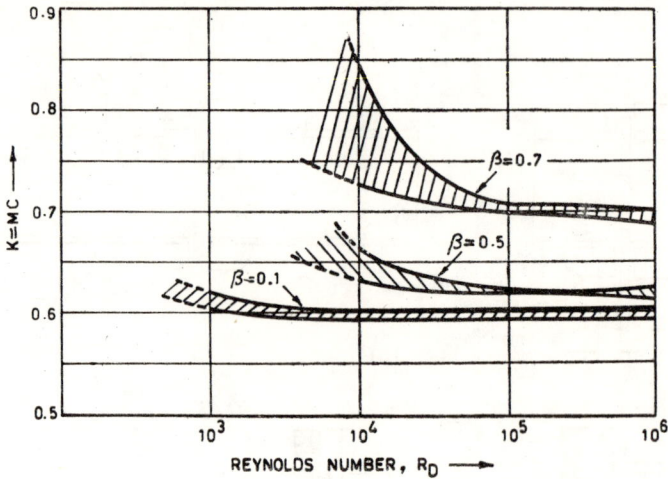

Fig. 7.5 Dependence of flow coefficient on Reynolds number R_D for various values of β for flat plate orifices

Flow Nozzles

The recommended proportions for the flow nozzles are given in Fig. 7.4 (b). The approach curve must be proportioned to prevent separation between flow and the wall, and parallel section is used to ensure that the flow fills the throat. Usually pipe taps are used. Figure 7.6 illustrates the dependence of discharge coefficient on Reynolds number for various values of β for a long radius nozzle.

Venturi Tube

The recommended proportions of a standard venturi tube are shown in Fig. 7.4 (c). Note that the pressure taps are connected to manifolds which surround the upstream and throat portions of the tube. These manifolds receive a sampling of pressure all around the sections so that a good average value is obtained. The discharge coefficient as a function of Reynolds number for a venturi is shown in Fig. 7.7 with the tolerance indicated by the dotted lines.

When the compressible fluids are to be metered, the expansion coefficient should be calculated from the relations given earlier and the flow rate then calculated. The variation of expansion coefficient, Y, for orifices, nozzles, and venturi tubes as a function of pressure differential for $\gamma = 1.4$ is given in Fig. 7.8. Note the linear dependence of Y on ratio p_2/p_1 for whole rang of β values for orifices and practically so for nozzles and ventures. The measurement of flow rates by the obstruction meters reduces to the measurement of pressure differential.

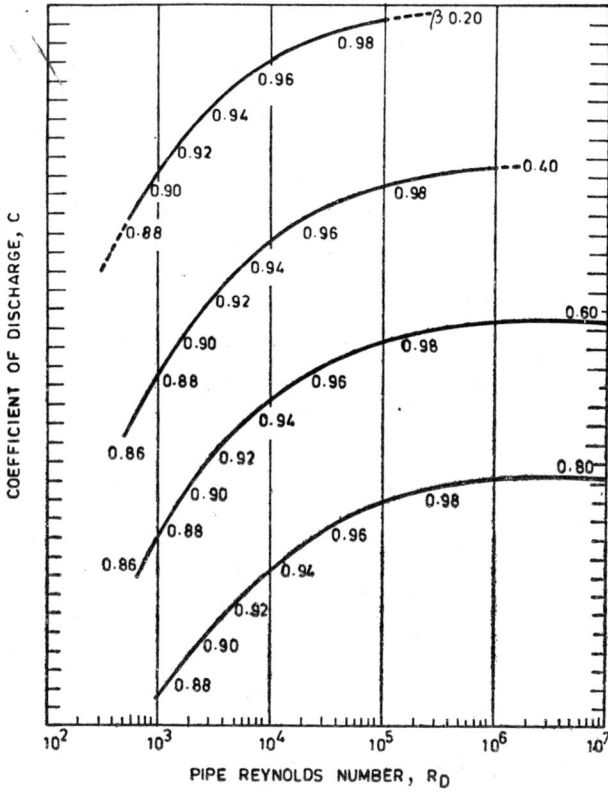

Fig. 7.6 Discharge coefficients for long radius nozzles for various values

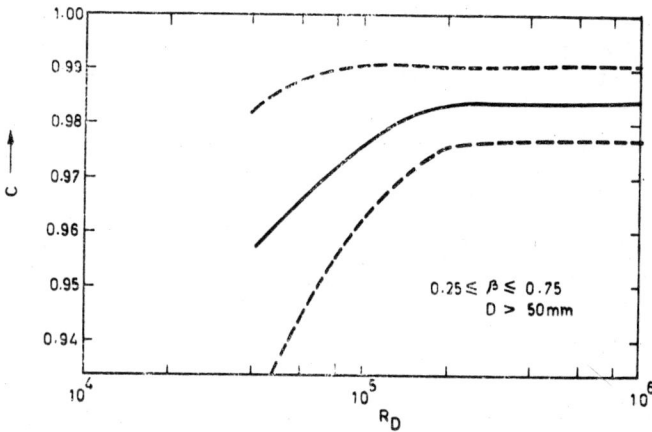

Fig. 7.7 Discharge coefficient for venturi tubes as a function of
Reynold's number

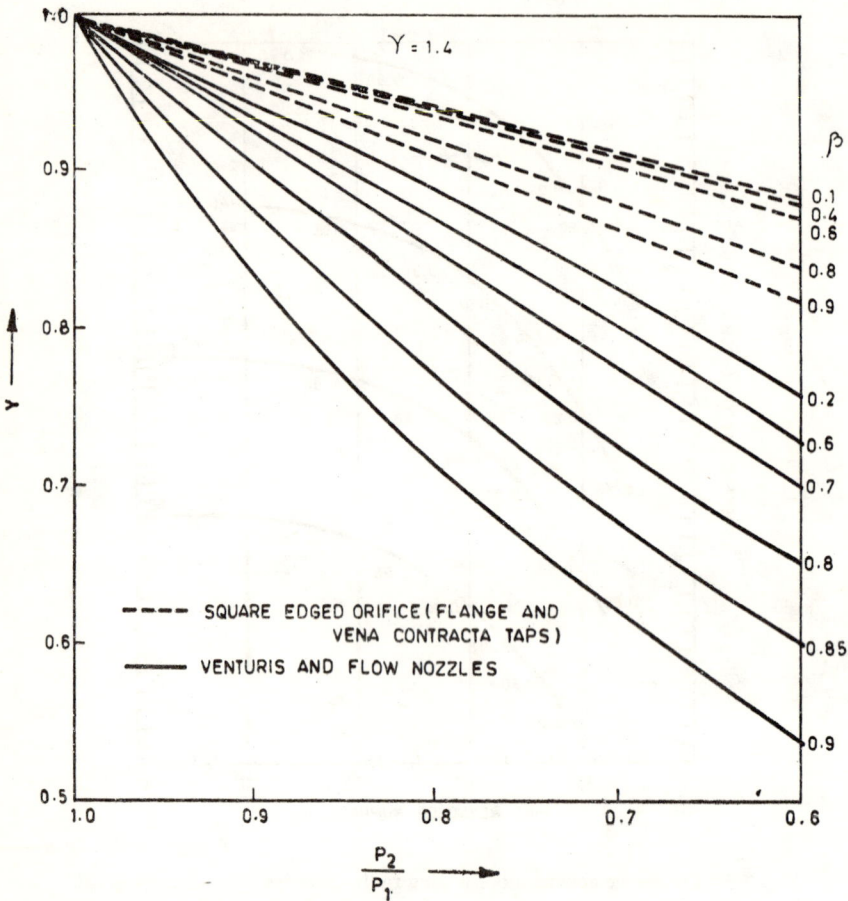

Fig. 7.8   Variation of expansion coefficient with pressure ratio

MEASUREMENT OF PRESSURE DIFFERENTIAL

The determination of flow rate using obstruction meters involves the measurement of pressure differential. Many meters are available, for example, mercury float flow meter, Ledoux-bell meter, ring-balance meter, aneroid meter, force-balance type meter, inductance bridge flow meter, etc., wherein pressures from the upstream and downstream sides of the primary element are fed in simultaneously. Some of the meters extract the square roots, like Ledoux-meter, so that the output is linear, in others square root chart paper or scale, like in ring-balance meter, is used for obtaining linear output. One such meter, the ring-balance meter, is discussed in detail here.

Ring-balance Meter

The ring-balance meter is a radial torque meter which uses a hollow ring body to convert the pressure differential generated by a differential medium, or by a difference in static pressure, into a rotation which is transmitted to the recorder or indicator.

Figure 7.9 shows the ring assembly in a typical flow-measuring system. Figure 7.9 (a) indicates the position when pressure differential exists while Fig. 7.9 (b) indicates the position of ring body at pressure equalisation or zero flow condition.

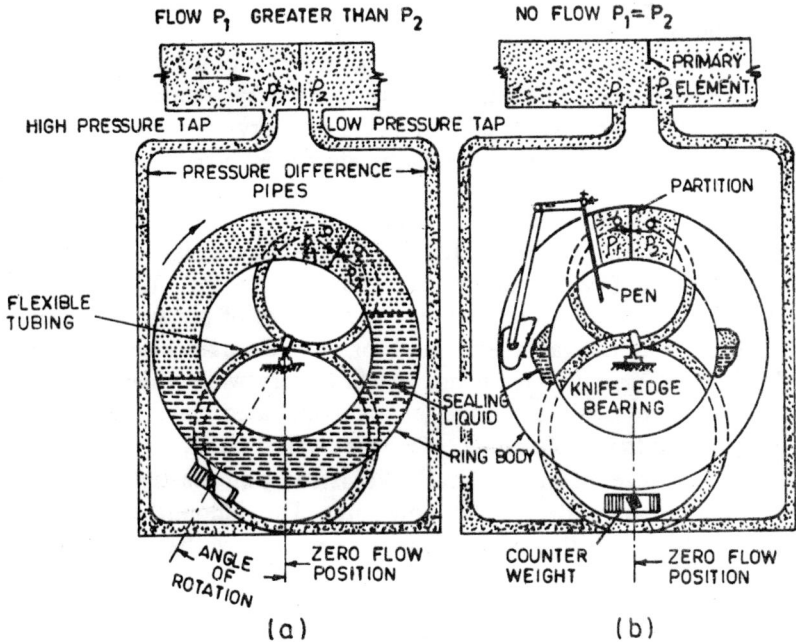

Fig. 7.9   Ring balance meter: (a) Flow on and (b) No flow

The ring assembly is mounted on a knife-edge bearing, which permits the rotation about the axis of the ring. The ring is divided into two pressure compartments by the baffle at the top, and by the sealing liquid usually mercury which fills the lower part of the ring. The two ring compartments thus formed are connected to the differential pressure pipes by means of flexible tubing to permit the ring to rotate freely under the action of the difference in pressure in the compartment.

The ring rotation is transmitted to the recording pen or indicating pointer by a linkage. When the meter is used to register flow rate, a square root chart or a scale is employed. In some cases the ring rotation is transmitted to the pen through a cam and the cam follower. The cam follower imparts a movement to the pen which is directly proportional to the flow rate.

Ring torque is a function of differential pressure acting on the baffle. The torque is resisted by an external calibration weight, rigidly attached to the bottom of the ring body. The meter is in equilibrium when the ring torque is balanced by this counter weight.

Mercury or other sealing fluid exerts no force tending to rotate the ring, but acts only as a seal for the differential pressure in the two compartments. This is true because in the circular body all hydrostatic forces resulting from the deflection of the sealing liquid are directed normally to the containing circle, and therefore pass radially through the exact centre of rotation without producing meter torque. The ring-balance may contain two S-shaped, self-compensating tubes for relatively high static pressure or parallel tubes for lower pressure applications.

It should be noted that the flow measurements based on the measurement of pressure differential across a primary element like orifice, nozzle, venturi etc., essentially fall in the category of indirect methods, as measurement of several other parameters is involved.

VARIABLE AREA METERS

In obstruction meters, restriction in the channel is of fixed size and the pressure differential across it changes with the flow rate. The flow rate is proportional to the square-root of pressure differential. This is often a disadvantage because for the measurement of wide ranges of flow rates, pressure measuring system of very wide range is required. In other words, if the range is accommodated, the sensitivity is not uniform over the whole range. On the other hand, in variable-area meters, the size of restriction is adjusted by an amount necessary to keep the pressure differential constant. The amount of adjustment required is a function of flow rate. There are two basic types of area meters: piston type meter and rotameter.

Piston Type Meter

It is designed specifically for metering viscous liquids, such as hot tar, black liquor, etc., which are difficult to measure in any other way. It is installed directly in the pipeline and is usually equipped with an electric transmitter. If the rate of flow increases, the differential pressure across the metering plug tends to increase. This raises the plug and increases the port area in proportion to the rate of flow. The converse is true when the flow decreases. Two types of piston type meters are shown in Fig. 7.10 (a) and (b).

Rotameter

The rotameter consists of a tapered tube, mounted vertically with the tube diameter increasing upwards. The tube carries a metering float usually called bob or float. The direction of flow has to be vertically upwards at the meter. When the fluid is not flowing, the bob rests at the bottom of the tube and its maximum diameter is so selected that it blocks the small end of the tube almost completely. When the flow begins, the bob rises till the annular passage between the inner wall of the tapered tube and the periphery of the bob is large enough to allow all the flow coming through the pipe. The bob then comes to rest in dynamic equilibrium and its position in the tapered tube is a measure of the flow rate.

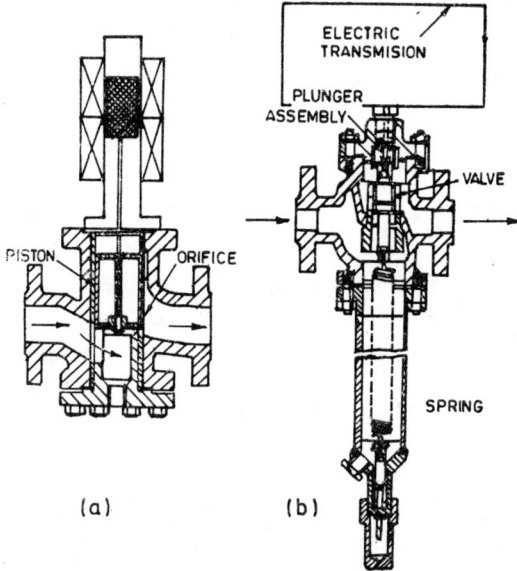

Fig. 7.10 (a) Piston type weight loaded area flow meter with electric transmission

(b) Piston type spring loaded area flow meter with electric transmission

Figure 7.11 shows a schematic of a rotameter. Its smaller end has a diameter D and its taper is given by an angle α. The maximum diameter of the bob is d and its volume V. The rotameter meters the flow of a fluid of density ρ_f. At the rest position in dynamic equilibrium, the float is under the action of following forces:

(i)   Weight of the float acting downward $= \rho_b v g$

(ii) Buoyancy force acting upward $= \rho_f v g$

(iii) Force due to pressure differential acting upward $= (p_1 - p_2) A_e$

(iv) Viscous forces, which are very small in magnitude, and have been neglected in the analysis.

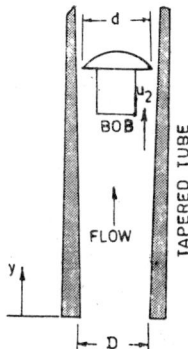

Fig. 7.11   Schematic of a rotameter

Here ρ_b is the density of the material of float and A_e is its effective projected area. At dynamic equilibrium, one has

$$(p_1 - p_2) \, A_e = (\rho_b - \rho_f) \, vg$$

Assuming an incompressible flow, the outgoing velocity of the fluid is expressed as

$$u_2^2 = \frac{2}{1 - (A_2/A_1)^2} \left(\frac{p_1 - p_2}{\rho_f} \right)$$

This equation is obtained from the Bernoulli equation and the continuity equation. A_1 is the in-area of the device and A_2 is the annular area between the wall of the tapered tube and the periphery of the bob. Therefore,

$$A_2 = [\pi/4] \, [(D + 2ay)^2 - d^2] \simeq \pi \, Day$$

where y is the vertical distance of the float from the entrance and usually the diameters $d \simeq D$. Substituting for $(p_1 - p_2)$ from the equation of the dynamic equilibrium, the velocity u_2 is given by

$$u_2 = \frac{1}{[1 - (A_2/A_1)^2]^{1/2}} \left[\frac{2vg}{A_e} \cdot \frac{\rho_b - \rho_f}{\rho_f} \right]^{1/2}$$

The actual rate of flow $\dot{Q}$ is thus,

$$\dot{Q} = c_d A u_2 = \frac{c_d A_2}{[1 - (A_2/A_1)^2]^{1/2}} \left[\frac{2vg}{A_e} \cdot \frac{\rho_b - \rho_f}{\rho_f} \right]^{1/2},$$

where the effect of drag force is included in the coefficient c_d, called the drag coefficient.

If the variation of c_d with the bob position is very small, and also the value of $\left(\dfrac{A_2}{A_1} \right)^2$ with bob position remains much smaller compared to unity, the flow rate $\dot{Q}$ can be expressed as

$$\dot{Q} = c'y \left(\frac{\rho_b}{\rho_f} - 1 \right)^{1/2}$$

where $c' = \dfrac{c_d \pi a}{4 \, [1 - (A_2/A_1)^2]^{1/2}} \left(\dfrac{2vg}{A_e} \right)^{1/2}$ is a constant.

Thus every float position, y, corresponds to one particular flow rate and no other. It is necessary to provide a reading or a linear calibration scale on the outer side of the tube. The flow can be determined by direct observation of the position of the metering float.

The equation of the mass flow rate $\dot{m}$ can be written as

$$\dot{m} = \rho_f \dot{Q} = c'y \, [\rho_f \, (\rho_b - \rho_f)]^{1/2}$$

The mass flow rate depends on the density of the fluid being metered. Any variations in the fluid density due to the temperature variation will result in variation in the mass flow rate. It is frequently advantageous to have a rotameter which gives an indication of flow independent of the change in the fluid density. By a proper choice of the material of the float, the mass flow rate can be made very insensitive to the density changes. The condition which satisfies this is $\rho_b = 2\rho_f$. Then the mass flow rate is given by

$$\dot{m} = c'y\rho_f = c'y\rho_b/2$$

By the use of a float constructed from a material satisfying the condition $\rho_b = 2\rho_f$, the meter itself compensate for density changes of the fluid due to temperature changes while metering the flow. The error in metering $\dot{m}$ under this condition is less than 0.2% for a fluid density variation of 5% from the above condition.

VELOCITY PROBES

When the description of a flow field is desired, both magnitude and direction of flow-velocity vector at various points in the field should be known. This is achieved by variety of pressure probes; measurement of flow velocity vector is carried out over a finite area due to the finite dimensions of probe, instead of at a 'point'. Thus, the direct measurement results in the values of average flow conditions.

The choice of a particular probe rests on many factors such as the type of information required, size, flow conditions, etc. Basically, pressure probes measure either of the two different pressures or some combination thereof. These are static p_s and total p_t pressures, such that

$$p_t = p_s + p_v$$

where p_v is the velocity pressure.

Static pressure, p_s, is the actual pressure of the fluid whether at rest or in motion. Velocity or dynamic pressure, p_v, is the pressure equivalent of the directed kinetic energy of the fluid if the fluid is considered as a continuum. For an incompressible fluid, it is given by

$$p_v = \rho \frac{v^2}{2g},$$

where ρ is the density of the fluid, and v is its velocity at a point where p_v is given. Total pressure, p_t, is the sum of static and dynamic pressures. It is also called as stagnation, or impact or pitot pressure.

Substituting for velocity pressure and rearranging, an expression for the flow velocity v is given as

$$v = \sqrt{\frac{2g\,(p_t - p_s)}{\rho}}$$

Therefore, the velocity may be determined simply by measuring the difference between total and static pressures.

The above equation holds good for incompressible fluid. For compressible flow, the above relation is suitably modified. Assume an ideal gas (compressible fluid) undergoing an isentropic process, then $p/\rho^\gamma = \text{const.}$ Thus

$$\left(\frac{p_t}{p_t} - \frac{p_s}{p_s}\right)_{comp} = \left(\frac{\gamma - 1}{\gamma}\right)\left(\frac{v^2}{2g}\right),$$

where subscript 'comp' means compressible. The relation can be expanded in terms of Mach number M defined as $M = (v/\gamma g R T)^{1/2}$
Thus

$$(p_t - p_s)_{comp} = \frac{\rho v^2}{2}\left(1 + \frac{M^2}{4} + (2 - \gamma)\frac{M^4}{24} + \ldots\right) \text{ for } M^2\frac{(\gamma - 1)}{2} < 1.$$

When $M \to 0$, the above relation reduces to that of incompressible fluid.

The effect of compressibility can usually be taken into account through the use of a correction factor c' thus

$$v = (1 - c')\sqrt{\frac{2g\,(p_t - p_s)}{\rho}},$$

Figure 7.12 gives the approximate velocity corrections for fluid compressibility based on air at atmospheric pressure.

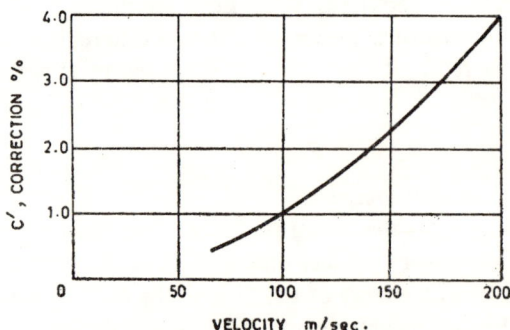

Fig. 7.12   Velocity correction for fluid compressibility

When the velocity is used to measure flow rate, a weighted velocity is to be obtained by measuring it at various points in the channel, or a multiplication factor may be determined by calibration, for a given Reynolds number.

Static Pressure measurement

The static pressure can be sensed in the following ways:

Wall Taps: Small holes can be drilled in the surface of the flow boundry in such a way that streamlines of the flow remain relatively undisturbed. The accuracy of measuring static pressure with wall taps is determined by the size and shape of the hole. It has been shown that for a smooth pipe with incompressible turbulent flow and a static pressure hole diameter of 1/10 of the pipe diameter, the static pressure error reaches about 1% of the mean dynamic pressure at a pipe Reynolds number of 2×10^5.

Static Tubes: The accuracy in static pressure measurement using static tubes depends on the position of the sensing holes with respect to the nose of the tube and main supporting stem. Streamlines next to the nose of the tube must be longer than those in the undisturbed flow, indicating an increase in the velocity. Acceleration effects thus caused by the nose tend to lower the tap pressure, while the stagnation effects caused by the stem tend to raise the tap pressure. The static tube characteristics are shown in Fig. 7.13. In the figure x_h is the position of taps from base of nose and x_s is the position of taps from centerline of stem; both are normalised with respect to the diameter of the tube.

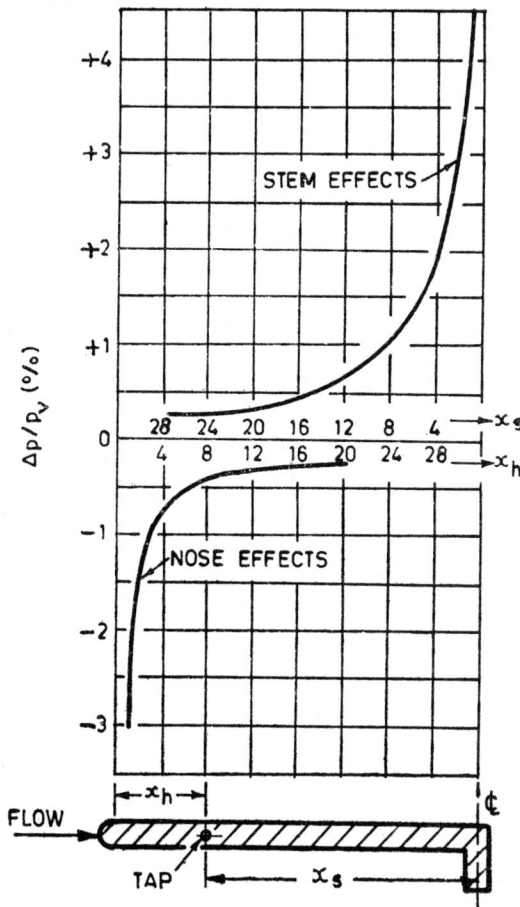

Fig. 7.13   Static tube characteristics

In a properly compensated tube, the acceleration and stagnation effects will just balance each other at the plane of the pressure holes. This is the principle of Prandtl-Pitot tube. Figure 7.14 shows the design of this tube which utilises eight square edged pressure holes (1 mm in diameter) placed 45° apart in a plane located 8 tube diameters downstream of the nose and 20 tube diameters upstream of the probe stem. The disc probe is another static pressure sensor that uses the compensation principle.

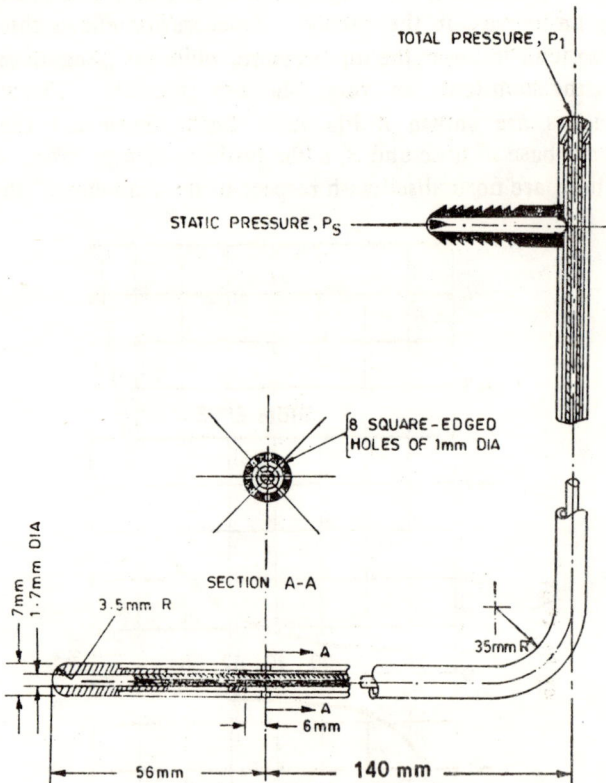

Fig. 7.14 The Prandtl-Pitot tube

Aerodynamic Probes

Cylindrical probes inserted normal to the flow fall under this category. The pressure distribution over the surface of a cylinder is well known and is shown in Fig. 7.15. The static taps are fixed at the critical angle. The critical angle is the angle at which static pressure occurs on the surface of the cylinder. In order to find the direction of flow normal to the cylinder, a cylindrical probe having two taps located in the same plane and separated by twice the critical angle are used. The probe is placed normal to the flow and is rotated about its axis.

The angular position of the probe is adjusted such that the two pressure taps, connected across a manometer, sense identical pressures. The flow direction now corresponds to the bisector of the angle between the two taps. The wedge probe is also used for this purpose.

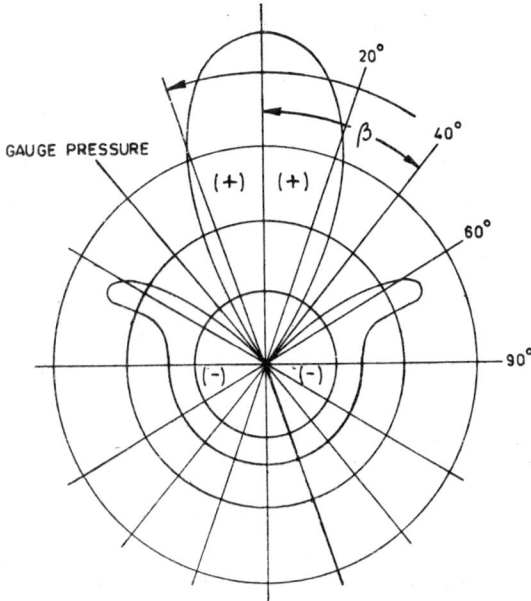

Fig. 7.15   Pressure distribution on surface of a cylinder
inserted normal to flow

Figure 7.16 illustrates the comparison of performances of cylindrical and wedge probes. It is obvious that the wedge probe has a less rapid change in tap pressure in the region of pressure taps than does the cylindrical probe. Unfortunately, the fragile apex of the wedge makes it a less robust instrument and favours the use of cylindrical probe in many applications. The static pressure probes are used for measuring static pressures in obstruction meters. They are required in velocity determinations to establish thermodynamic state points and are also useful in obtaining indications of flow direction.

Total Pressure Probes

The measurement of total or stagnation pressure is usually somewhat easier than the measurement of static pressure. By definition the total pressure can be sensed by bringing the flow to rest isoentropically. The stagnation pressure can be measured adequately with the classical Pitot tube. More often, a Pitot tube is provided with the static openings; one such using a compensation principle is the Prandtl-Pitot tube shown in Fig. 7.14. The stagnation pressure can also be sensed by holes located at stagnation points on aerodynamic bodies such as spheres and

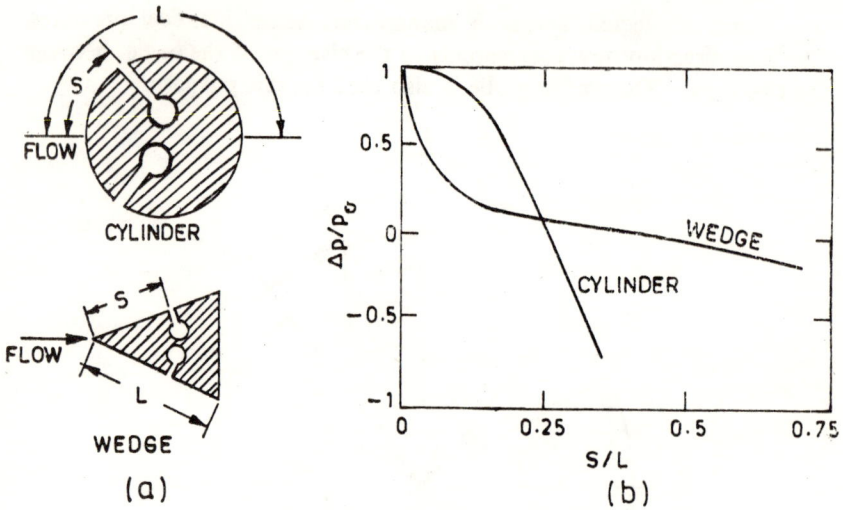

Fig. 7.16 (a) Cylinder and wedge shaped probes and
(b) Their performance characteristics

cylinders. In all these it is assumed that the fluid is brought to rest isoentropically in the vicinity of the tap. The departure from true stagnation pressure is indicated by Pitot coefficient c_p, where

$$c_p = \frac{p_{tI} - p_s}{p_v}$$

where p_{tI} is the indicated total pressure. The Pitot coefficient is unity under usual flow conditions. However, it is influenced by viscosity, probe geometry, misalignment with flow direction, etc.

The Kiel tube, designed to measure only stagnation pressure possesses a remarkable insensitivity to yaw angle (misalignment with respect to flow stream lines). It consists of an impact tube surrounded by a smoother venturi tube (Fig. 7.17).

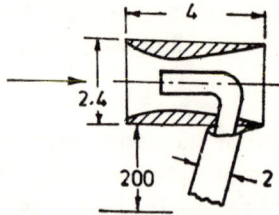

Fig. 7.17 The Kiel probe

A relatively cheaper probe having essentially similar performance characteristics as that of Kiel tube is obtained by employing a cylindrical duct, bevelled at each end rather than a streamlined venturi.

7.5   Special Methods

TURBINE TYPE FLOW METER

A turbine wheel is mounted axially on bearings inside the flow pipe. As the fluid flows, it causes the turbine wheel to rotate. A permanent magnet is installed inside the turbine body in such a way that it rotates with the wheel. A reluctance pick up produces voltage pulses as the turbine wheel rotates: one pulse per rotation of the wheel is produced. It is possible to design a turbine whose speed varies linearly with volumetric flow rate. Therefore, the pulse rate is proportional to the flow rate. At low flow rates, viscosity effects cause the response to be nonlinear. However, at the high flow rate the response is linearly related with the frequency of rotation of the turbine wheel.

In general, the volumetric flow rate $\dot{Q}$ is related to the pulse frequency f through a relation:

$$\dot{Q} = \frac{f}{K},$$

where flow coefficient K for the turbine flow meter can be obtained from the calibration curve. It depends on the design parameters of the meter, flow rate and kinematic viscosity of the fluid. For high flow rates, K is nearly constant for any given meter. The commercial turbine meters can handle flow rates in a wide range. The meter is driven by the flow itself. The pressure loss at maximum flow rate is around $5 \times 10^4 \ N/m^2$. The transient response of the meter is good. However bearing maintenance problems are always present.

ELECTROMAGNETIC FLOW METER

Electromagnetic flow meters are based on the Faraday principle of induction which states that the emf in a conductor moving through a magnetic field is proportional to its velocity. If a conductor of length l moves with a transverse velocity v across a magnetic flux density B, an emf e is developed. It is expressed as

$$e = Blv \times 10^{-8} \text{ volts}$$

where
$$e = \text{emf in volts}$$
$$B = \text{magnetic flux density}$$
$$l = \text{length in cm, and}$$
$$v = \text{velocity in cm/s.}$$

In practice the length of the conductor is equal to the inside diameter D of the pipe. The emf developed can be expressed in terms of the volumetric flow rate $\dot{Q} = (\pi D^2 v)/4$ as $e = (4B\dot{Q}/\pi D) \ 10^{-8} \ volts$. For less conducting fluid, like water, a non-conducting pipe-line is used and the electrodes are flushed with it so that they are in contact with the fluid. The pipe is of non-magnetic material as to allow the magnetic field to penetrate. The flow is usually assumed to be uniform, however, the result holds good for all symmetrical flows but then the average flow velocity is indicated.

For highly conducting fluids, like mercury, the metallic pipe (stainless steel is used as the pipe metal) is not very effective in short circuiting the voltage induced in the fluid flow. This is because the electrical conductivity of mercury is much higher than that of stainless-steel. Both d.c. and a.c. magnetic fields can be used; the d.c. magnetic field is preferred for metallic flow as the output is high and hence no amplification is required. While for liquids which can be polarised, an a.c. field is used and further amplification is possible which is usually required for less conducting fluids.

The main limitation of electromagnetic flow meters is the conductivity of the fluid; it must be sufficiently high so that the external circuitry is not an excessive load. However, it possesses the advantages of complete absence of any obstruction in the pipe, ability to measure reverse flows, insensitivity to viscosity, density and any other flow disturbance so long the flow is symmetrical, wider linear range and good transient response.

SONIC FLOW METER

This flow meter is based on the principle of addition of velocities—the one of ultrasonic vectorially added to that of flow field and the resultant velocity is measured. In this flow meter, pressure waves of frequency around 10 MHz are sent from an ultrasonic source either a piezoelectric or magneto-strictive transducer located externally on the pipe and picked up by the receiver on the opposite side of the transducer. In fact one measures the time delay in this method. In another variant of this method, the beat frequency is measured. The signal travelling through the flowing medium suffers a frequency change, which is proportional to the flow velocity. The frequency of this signal is compared with that of the reference signal.

In another approach one senses the deflection of ultrasonic beam propagated transversely to the flow. Two receivers longitudinally displaced are mounted on the opposite side of the tube. With no flow, the two receivers pick up the signals of identical strengths. With flow, the signal strength at one receiver will increase and at the other will decrease. The ratio of these signals is linearly related with the flow rate.

The ultrasonic flow meters have the advantages of not using any flow obstruction, insensitivity to viscosity, temperature and density variations. They are suited for the flow measurement of liquids. They are however more expensive.

HOT WIRE ANEMOMETER

Hot wire anemometer is a device which is most often used in research applications to study varying flow conditions. This is used in two basic modes: constant temperature mode, and constant current mode.

A thin wire of resistance R carrying a current i is exposed to the flow field. The wire attains an equilibrium temperature when the Joule heating i^2R is balanced by the convective heat loss from the wire. The convective heat loss is given by $hA(T_w - T_f)$, where h is the convective film coefficient, A is the heat transfer area, T_w and T_f are the temperatures of the wire and the flowing fluid respectively. The convective film coefficient is mainly a function of flow velocity for a given fluid density. The functional relationship between h and flow velocity u valid over a range of velocities is of the form

$$h = a + b \sqrt{u}.$$

Therefore as the flow velocity changes, the convective heat loss changes resulting in a variation in wire temperature. In constant temperature mode, the temperature of the wire is kept constant by continuously varying the current flowing through it. The current flowing through the wire is taken as a measure of flow velocity. The temperature is sensed by measuring the resistance of the wire by Wheatstone bridge or its modifications. In constant current mode, the Joule heat i^2R is essentially kept constant. As the flow velocity changes, the temperature of the wire adjusts itself until the equilibrium is reached. The temperature of wire is then taken as the measure of flow velocity.

A common hot wire anemometer consists of a thin wire of about $8\mu m$ diameter and 1 mm long, having a resistance of approximately 1 ohm and supported on a non-conducting structure (Fig. 7.18 a). The wire is usually made of platinum, nickel or tungsten. The probe is effectively of a point size. The main disadvantages of the hot wire anemometer lie in the limited strength, calibration changes caused by the accumulation of impurities on the wire, vibration of wire resulting in quick damage and flutter effects. Unless the flow field is clean, its use is not recommended. It has, therefore been used mainly in gases.

In liquids, a variation of it, often called thin film anemometer is used. A thin film of platinum is coated on a pyrex glass wedge and the

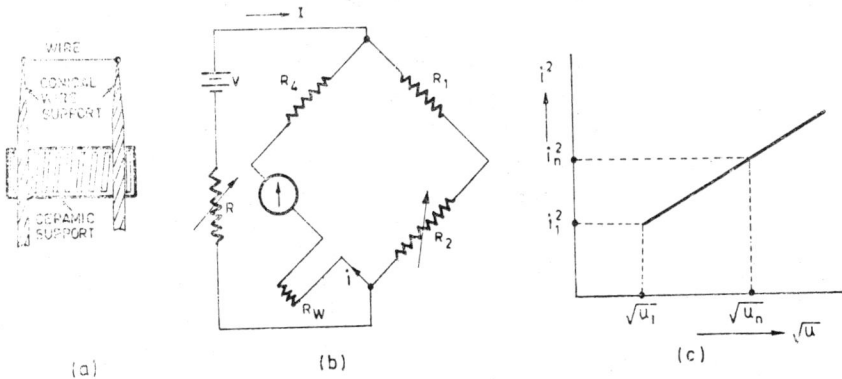

Fig. 7.18 (a) Hot-wire anemometer probe, (b) Wheatstone circuit for constant temperature mode, and (c) Calibration curve

connections are taken through heavy silver plates. It is very robust, easy to clean and can be used at higher temperatures.

It has been mentioned earlier that an equilibrium temperature is attained when the Joule heating in the wire is just balanced by the convective heat loss. Therefore at equilibrium,

$$i^2 R_w = K_c h A(T_w - T_f)$$

where R_w is the resistance of the probe, and K_c is a conversion factor from thermal to electrical quantities. Substituting for h, one obtains

$$i^2 R_w = c_0 + c_l \sqrt{u},$$

where $c_0 = K_c A(T_w - T_f)a$, and $c_l = K_c A(T_w - T_f)b$.

Therefore, a relation between i^2 and $\sqrt{u}$ is linear. This functional relationship is obtained by the process of calibration.

Constant Temperature Mode

The constant temperature mode of operation is often used for the measurement of average or steady flow velocities, as the frequency response is restricted by the associated electronics. At higher frequencies the system becomes unstable and the wire temperature oscillates. But this mode of operations has the advantage that the wire is inherently protected against burnout.

A given hot-wire probe is to be calibrated in a fluid in which it is expected to be used later. The calibration is carried out by exposing the probe to known velocities (measured accurately by some other means) and recording its output over a range of velocities.

Figure 7.18(b) shows a schematic circuit diagram of the hot-wire anemometer when operated in a constant temperature mode. The resistances R_1, R_2, R_w and R_4 forming a bridge circuit are of the order of 1 to 20 ohms each. The resistance R is of the order of 2kΩ. Therefore, the current flowing through R_w remains essentially constant even when R_w changes. The probe is calibrated by exposing it to some flow velocity say u_l, and the current i flowing through R_w is so adjusted by varying as to provide adequate sensitivity without any danger of burnout. The resistance of the probe will settle to some value depending on the flow velocity. The bridge is now balanced by varying the resistance R_2. Let i_l be the current flowing through R_w. The hot-wire probe will attain a certain temperature which is to be kept constant when it is exposed to other flow velocities. The first point on the calibration curve thus is $(i_1^2, \sqrt{u_1})$. The hot-wire probe is now exposed to another velocity u_2. This would change the temperature of the wire, and hence its resistance resulting in an unbalanced bridge. The bridge is restored by changing the value of current to i_2 by means of R. This restores the resistance to R_w, and thereby the wire temperature. This gives the second point on the calibration curve. The procedure is repeated for as many velocities as desired. During the whole calibration procedure the resistance R_2 is never changed. The calibration curve between i^2 and $\sqrt{u}$ is obtained by the method of least squares from the measured data. Once calibrated, the probe can be used to measure unknown velocities by adjusting the value of R until the bridge balance is achieved. The velocity u corresponding to the current i at the balance is read off from the calibration curve. It however assumes that the temperature and pressure during the measurement are at the same values prevailing during calibration of the probe.

Instead of measuring current flowing through R_w by a meter, a standard resistance can be put in series with R_w and the voltage drop across it is potentiometrically compared to obtain the value of i. This mode of operation can also be adapted to the measurement of transient flows. With reference to the measurement of transient flows, constant current mode of operation of the probe is now discussed.

Constant Current Mode

Figure 7.19 (a) shows a schematic circuit diagram of the hot-wire anemometer operation in a constant current mode. The resistance R is of the order of $2K\Omega$ while that of the probe is about 1Ω. Any variation in R_w due to flow changes will not change the value of current which remains practically constant. Time constants of the order of 1 m sec. can be achieved with 2.5 μm diameter wires of platinum or tungsten operating in air. The constant current mode is, therefore, suited for the study of fluctuating flow fields. However large variations in the flow rate may lead to wire burnout. When the probe is exposed to a fluctuating component of velocity field of the type $u(t) = u_0 + v(t)$, where u_0 is an average velocity component and $v(t)$ is the fluctuating component around u_0, the output e is also fluctuating. The variation of velocity with time and consequently of output e with time is shown in Fig. 7.19 (b) and (c) respectively. The transient response of the system is first order and is given by

$$\frac{e}{v}(D) = \frac{k}{\tau D + 1}$$

where e is the output voltage, k is the static sensitivity and τ is the time constant. The time constant of the hot-wire anemometers usually is in the range of 1 ms and hence their flat frequency response is less than 160 Hz. The probe certainly would not respond to the frequencies of the order of 50 KHz encountered in the study of turbulent flows. This limitation is overcome by electrical dynamic compensation which raises the frequency response to MHz range.

The compensation network reduces the magnitude of the signal considerably. But the main drawback of the compensation technique is that the correct compensation depends on the value of τ, the time constant of the fluctuating flow field which is not known and varies with flow conditions. This difficulty is overcome by superposing a square wave current

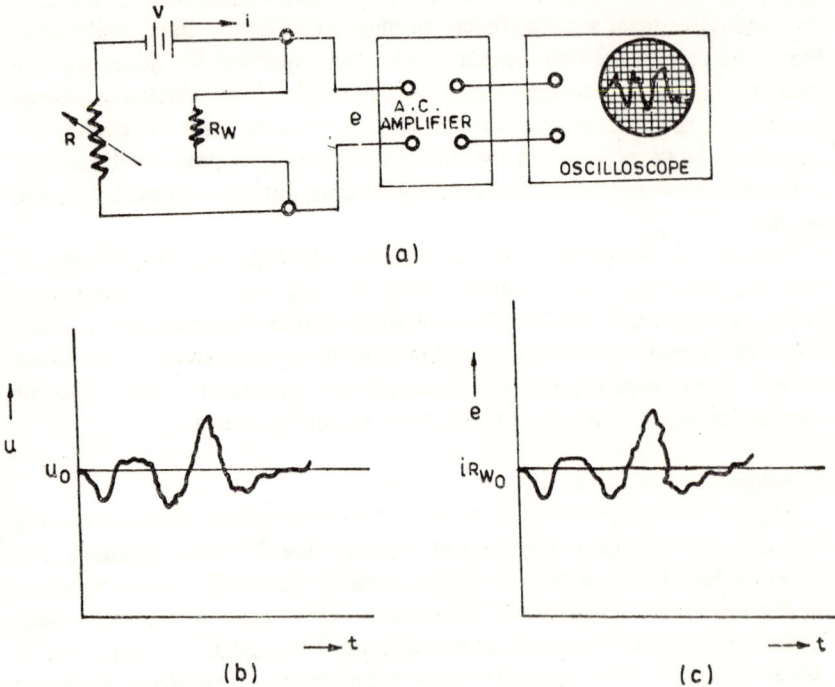

Fig. 7.19 (a) Hot-wire anemometer circuit-constant current mode
          (b) Fluctuations in flow-velocity, and
          (c) Fluctuations in output

on the d.c. current through the hot-wire while it is exposed to the flow to be studied. The correctness of compensation can be judged by the degree to which output voltage corresponds to the square wave. The adjustment to the compensation network is done while the hot-wire is exposed to flow.

VORTEX SHEDDING TECHNIQUE

Vortex shedding is a phenomenon that can occur when a fluid flows past a bluff or non-streamlined body. Generally, the flow does not follow the shape of the obstruction on the downstream side but separates from its surface, causing eddies to form. They grow in size until too large to remain attached to the surface. They then break away; the frequency of downstream shedding is determined by the flow rate.

The direct relationship between vortex shedding and flow rate makes the phenomenon of vortex shedding pertinent for flow metering. The method is extremely simple as it requires only the counting of shed vortices, called Karman vortices, to establish total flow. A flow meter based on the phenomenon of vortex shedding is capable of metering liquids and gases in a temperature range from 200°C to 300°C. The linearity of the instrument over a flow range of 100 : 1 is 5%.

7.6 Measurement of mass flow rate

Some applications require the knowledge of mass flow rate rather than volume flow rate, for example, range capability of an aircraft or a rocket is determined by mass of the fuel rather than volume. Process industries, in particular chemical industries, often require the knowledge of mass flow.

There are two basic approaches for metering mass. The first involves the measurement of volume flow rate with a simultaneous measurement of density. Pressure differential meters give signals proportional to ρv_{av}^2 which is multiplied by ρ and square root extracted to give mass flow rate by the processor, while the velocity flow meters give only v_{av} to which the value of density is multiplied. In the second approach some methods are used which are inherently sensitive to mass flow alone, for example; heat transfer, angular momentum transfer, Coriolis acceleration, gyroscopic action, shock waves, pulsating flow, rotating Pitots and vibrating Pitots. The measurement of mass flow requires complex instrumentation. Several designs of mass flow meters are based on the angular momentum transfer. In one such instrument the fluid is directed axially through an impeller rotating at constant speed. The torque required to drive the impeller is a function of the mass of fluid passing through and is given by

$$T = r^2 \omega \dot{m}$$

where T is the torque, r radius at the outlet, ω the angular velocity and $\dot{m}$ is the mass flow rate through the impeller. The torque is, therefore, directly related to the mass flow rate. It should be noted that the torque will not be zero when $\dot{m} = 0$ because of the frictional effects. Further viscosity variations would also cause the torque to vary.

A variation of this approach is to drive the impeller at a constant torque with some sort of slip clutch. The impeller speed is then a measure of mass flow rate, i.e.,

$$\omega = \frac{T}{r^2 \dot{m}}$$

The relationship between ω and $\dot{m}$ is inverse but the measurement of ω is far more easier compared to that of T.

7.7 Flow Visualisation Methods

The hot-wire anemometer is used for point by point mapping of the flow field. Its insertion in the flow field may disturb its characteristics.

Laser anemometer discussed in Chapter 9 is also used for point by point mapping of the field but is of non-contact type and possesses a very large dynamic range. On the other hand the flow visualisation techniques are whole field methods and provide very useful information about the flow conditions. In some cases quantitative information can also be extracted. They are divided in the following groups:

  (i) Shadowgraphy,
  (ii) Schlieren techniques, and
  (iii) Interference Methods.

These methods work due to the fact that the pressure or the flow field accordingly changes the refractive index of the medium and these refractive index variations either refract light waves or introduce a path difference when the light waves propagate through the medium.

These techniques are also used in the field of combustion where the variations in the refractive index are brought about mainly by temperature. The optical effects seen are due to refractive index variations which in turn are to be related with other parameters. These techniques are not used for flow metering rather for visualising the flow field past models in wind tunnel studies. Flow causes density variations, and consequently the refractive index variations. The variation of refractive index with density can be expressed as

$$n = 1 + \beta \frac{\rho}{\rho_s}$$

where β is a constant ($\beta = 0.000293$ for air) and ρ_s is the reference density taken at standard conditions. Usually $\beta\ (\rho/\rho_s) \ll 1$. The path of a ray in a medium with refractive index variations is to be studied. It is assumed that the medium can be thought of made up of planes of constant refractive indices. These planes are assumed to lie in the $x - z$ plane. The refractive index varies along the y direction. The problem therefore reduces to two-dimensional analysis. Due to the variation of refractive index along y direction, the ray follows a

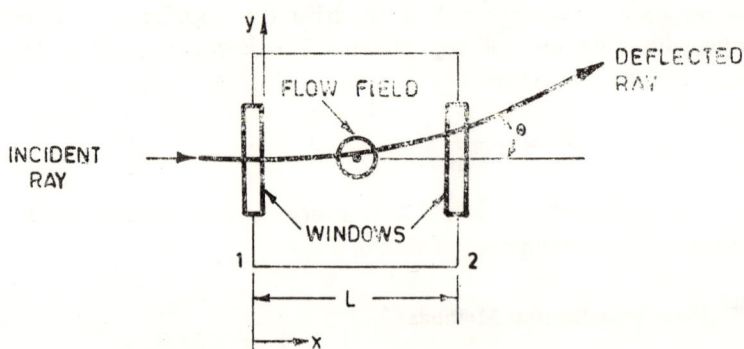

Fig. 7.20 Ray propagation in inhomogeneous medium

curved path in $x - y$ plane. The radius of curvature, R, of the ray path at a point (x, y) is

$$R = \frac{[1 + (dy/dx)^2]^{1/2}}{\dfrac{d^2y}{dx^2}}$$

The differential equation of the ray path is

$$\frac{(d^2y/dx^2)}{[1 + (dy/dx)^2]^{1/2}} = -\frac{1}{n}\frac{dn}{dy}.$$

where $\dfrac{dn}{dy}$ is the refractive index gradient.

It is now assumed that the $x -$ axis is along the incident beam. Therefore initially $(dy/dx) = 0$. Therefore, any deviation in the ray path will be due to refractive index variations. Furthermore (dy/dx) for air is very small, and hence $(dy/dx)^2 \ll 1$. Therefore, the equation of ray path is

$$\frac{d^2y}{dx^2} = -\frac{1}{n}\frac{dn}{dy}.$$

On integration, we have

$$\left[\frac{dy}{dx}\right]_1^2 = -\int_1^2 \frac{1}{n}\left(\frac{dn}{dy}\right) dx$$

or

$$\theta = -\int_1^2 \frac{1}{n}\left(\frac{dn}{dy}\right) dx$$

where $\theta \left(= \dfrac{dy}{dx}\bigg|_2 \right)$ is the angle of deflection of the emergent ray. The integration on the RHS is along the ray path; any variation along x direction implies a variation along y direction. But in practice the deflections are small, and hence the integration is carried along the x axis. Therefore, the angle of deflection θ is given by

$$\theta = \frac{L}{n}\left(\frac{dn}{dy}\right),$$

where L is the width of flow. The angle of deflection of a ray is directly proportional to the refractive index gradient in the flow. This is the basic optical effect on which shadowgraphy and Schlieren technique work. The angle θ can be expressed in terms of density gradients as

$$\theta = \frac{L\beta}{n\rho_s}\left(\frac{d\rho}{dy}\right).$$

It may therefore be noted that the deflection of a light ray is a measure of average density gradient integrated over the $x -$ coordinates. The analysis can be extended to three dimensions by including the refractive index variation along z direction. The deflection will now lie in $x - z$

plane. Therefore, the flow visualisation methods usually indicate refractive index variations in two dimensions only and will average the variations in the third dimension.

SHADOWGRAPHY

Figure 7.21 shows a schematic to view the flow field. A beam of collimated light illuminates the test space, and the observation is made on a screen in transmitted light. In the absence of flow, the screen appears uniformly illuminates. With the field 'on', various rays will suffer different deflections resulting in a non-uniform illumination at the screen. The displacement Δy at the screen will be

$$\Delta y = l\theta$$

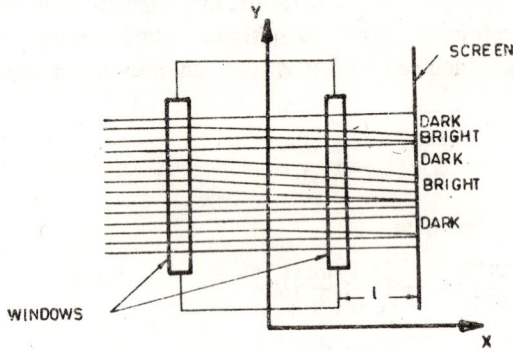

Fig. 7.21 Variation of intensity at the screen due to density gradients in the flow field

The irradiance at any point on the screen will depend on the relative deflection of the light rays at the screen, i.e.

$$\frac{d\theta}{dy} = \frac{d^2n}{dy^2} = \frac{d^2\rho}{dy^2}$$

The irradiance on the screen depends on the second derivative of the refractive index or density. It is almost fruitless to evaluate the refractive index or density distribution in the field from the shadowgraph. However, this is a powerful technique to see these variations and study them qualitatively.

Shadowgraphy can be carried out in divergent beam providing capability to cover larger test space, and variable geometric sensitivity. The shadow method has been utilised for the study of turbulent flow, formation and location of shock wave, etc.

SCHLIEREN TECHNIQUE

Consider a schematic shown in Fig. 7.22. A beam of light from a point source collimated by lens L_1 passes through the test section. This beam is focussed by a lens L_2 at a plane p_f where a knife edge is inserted. Another lens L_3 images the test section at plane p_l. In the absence of flow, the beam passes collimated and is focussed to a point at plane p_f. Since the image of a point source is also a point geometrically, the irradiance at the plane p_l will be either maximum or minimum depending on the knife-edge positions. However, when the flow is switched on, the rays suffer deflections and are no longer focussed to a point but scatter on a larger patch and in a nonuniform fashion. When the knife edge is inserted in the plane p_f the rays from various regions depending on deflection are obstructed resulting in a non-uniform intensity distribution. This effect is known as *Schlieren effect*. In practice, a slit source is used instead of a point source and the knife edge is moved parallel to the width of the slit as shown in Fig. 7.22.

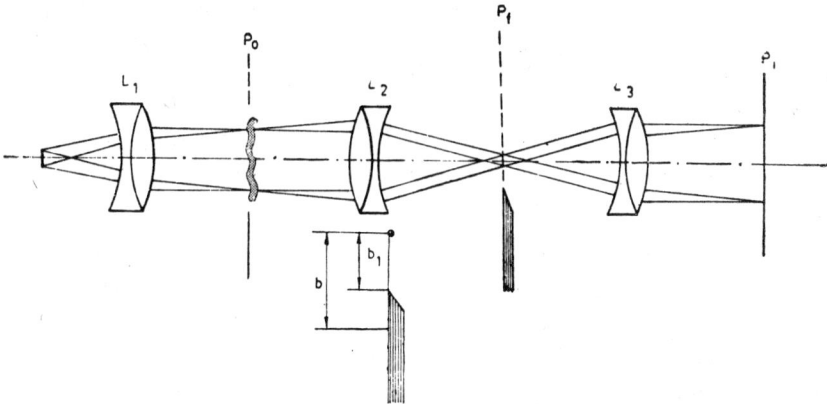

Fig. 7.22   Schematic of a Schlieren method

In the absence of flow, the irradiance at p_l decreases uniformly as the knife edge is inserted in and finally becomes zero when the image of the slit is completely blocked off. Therefore, the irradiance at p_l is proportional to the width of the slit that is not blocked. Let the width of image of the slit at the plane p_f be 'b' of which a width b_1 is not intercepted by the knife edge. Therefore, the irradiance distribution I, at this plane is proportional to b_1. With the flow 'on', a ray will be deviated by an angle θ. This deviation will displace its earlier position at p_l by

$$\Delta y = f\theta$$

where f is the focal length of the lens L_2. The irradiance variation ΔI over the uniform irradiance due to b_1 are due to the displacement Δy. Therefore the contrast c in the plane p_f is

$$c = \frac{\Delta I}{I} = \frac{\Delta y}{b_1} = \frac{f\theta}{b_1} = \frac{f}{b_1}\frac{L}{n}\left(\frac{dn}{dy}\right).$$

For air, it can be expressed as

$$c = \frac{f}{b_1} \frac{L}{n_1} \frac{\beta}{\rho_s} \left(\frac{d\rho}{dy}\right).$$

The contrast in the image of the test section is directly proportional to the refractive index or density gradients. The contrast can be enhanced by decreasing b_1, i.e. by intercepting more light resulting in the decrease in irradiance. The irradiance of the image can not be reduced indefinitely and hence a compromise between contrast and irradiance should be struck.

The use of slit over point source gives an advantage of more light but the knife-edge should always be aligned parallel to the slit. Further the component of deflection perpendicular to the knife-edge only produces Schlieren effect.

The important feature between shadowgraphy and Schlieren method is that the former senses the second derivative, while the latter the first derivative of refractive index or density. It is relatively easier to obtain quantitative information from Schlieren photographs. It will be shown that interferometry provides direct mapping of refractive index and hence is best suited for quantitative analysis. However, Schlieren technique has been used for the study of turbulent flows, boundary-layer phenomena, etc.

INTERFEROMETRIC TECHNIQUES

In spite of the fact that there are a variety of interferometers with certain advantages, Mach-Zehnder interferometer is still an unsurpassed choice of researchers. However, it is an exceedingly difficult device to align especially for a relatively longer period. It also suffers from the disadvantage of a direct and serious influence of external factors like vibration etc, . The beam from a source is collimated and split into two beams with the help of a beam splitter B_1 (Fig. 7.23). The choice of the initial collimating system is governed by the size of the field it is designed to offer. The two beams traverse at right angles to each other, folded by mirrors M_1 and M_2 and are re-united at the beam splitter B_2, where the interference effects are observed. The test section with viewing ports is in one arm of the interferometer, while the other arm has only the compensating plates. The interferometer is normally compensated for path. White light source then can be used for interference.

At least five independent motions are required for alignment and usually mirrors M_1 and M_2 are equipped with these. The interferometer may be adjusted either to have a uniform field or straight line fringes; the later is often preferred as the initial state of the system. These fringes are due to the interference between two plane wave-fronts which enclose a small angle between them. The orientation of fringes can be changed by adjusting the mirror alignment. The fringes are the loci of constant optical path. Therefore, if any variation in the test

COMPENSATING FILTERS

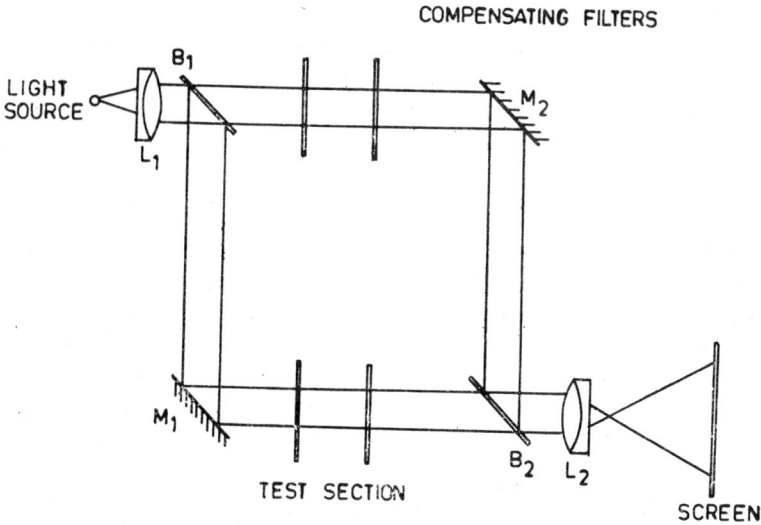

Fig. 7.23 Schematic of Mach-Zehnder interferometer

field occurs, it consequently changes the optical path and thus the shape and orientation of the fringes. A quick look at the interferogram provides information regarding the regions of constant optical path changes.

The path change Δ introduced in the beam as it passes through the test section of length L with the flow 'on' is

$$\Delta = L (n - n_0)$$

where n_0 is the refractive index in the reference path. Expressing refractive index in terms of density of the gas, the path difference Δ is expressed as

$$\Delta = L\beta \frac{\rho - \rho_s}{\rho_s}$$

The additional path introduced due to flow in the test section will cause the fringe to shift. Each fringe shift corresponds to a path change of one wavelength. Therefore, the number m of the fringes shifted is given by

$$m = \frac{\Delta}{\lambda} = \frac{L (n - n_0)}{\lambda} = L\beta \frac{\rho - \rho_s}{\rho_s \lambda}$$

where λ is the wavelength of light. When $n = n_0$ one obtains zero order fringe and therefore, the order of fringe is measured from zero order. The zero order fringe can be easily located using white light fringe pattern when the interferometer is compensated.

Another interesting interferometer which should find applications for flow visualisation is a triangular interferometer. This is insensitive to vibrations and environmental conditions like pressure and temperature changes because both beams follow the same path and are influenced equally. The schematic of the interferometer is shown in Fig. 7.24. The only disadvantage of the interferometer is that the reference beam also passes through the test section. However, the reference beam is focussed in the test space thereby reducing the influence of object changes on it. The interferometer uses interference between two spherical wave fronts rather than collimated wave. A very compact shearing interferometer using a Wollaston prism is commercially available. The density or temperature slope fringes are displayed.

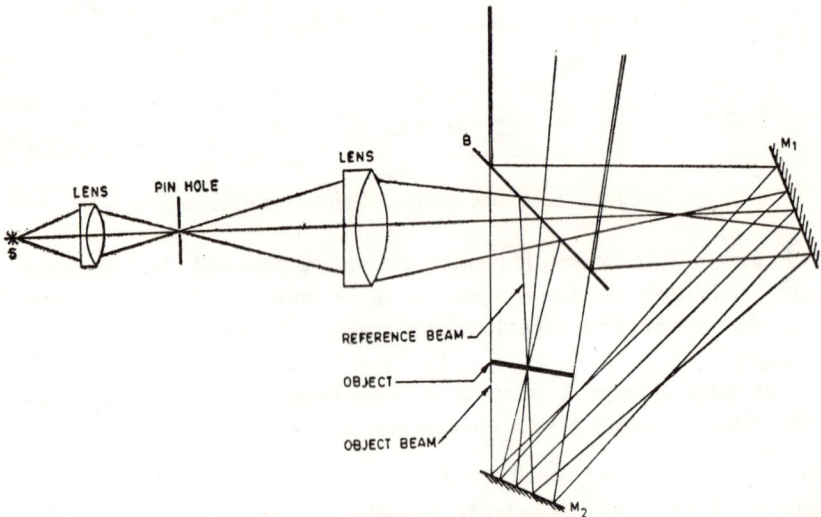

Fig. 7.24 The cyclic interferometer

The advantage of the interferometric method is that it directly gives the refractive index or density changes. These changes, however, have been integrated over the length of the test section. The interferometric method finds applications to a wide range of flow conditions from the low velocity ($\simeq 0.3$ m/s) flow in the free-convection boundary layers to the shock wave phenomena in supersonic flow.

Exercises

1. Nitrogen at a pressure of 2 N/m² and temperature of 35°C is being metered by a positive displacement meter. The uncertainties in the gas pressure and temperature measurement are $\pm 2 \times 10^{-4}$ N/m² and ± 0.5°C, respectively. The meter has been calibrated so that it indicates volumetric flow with an accuracy of $\pm 0.5\%$ from 30 to 100 m³/s. Calculate the uncertainty in the mass flow rate at the given pressure and temperature.

2. Analyse the error in the flow rate measurement caused by the thermal expansion of an orifice plate.

3. An obstruction meter is used for the measurement of the flow of moist air at low velocities. Suppose the flow rate is calculated taking the density of dry air at 30°C. Plot the error in flow rate as a function of relative humidity.

4. Linear response of a rotameter is obtained when the condition $\alpha y \gg D$ and $d \simeq D$ are met. Calculate the error introduced in the flow rate when $\alpha y = 10D$.

5. A rotameter is to be designed to measure a maximum flow rate of 40 lpm of water at 20°C. The bob is 25 mm in diameter and has a total volume of 15 cm³. It is made of a meterial having a density twice that of water at 20°C. The total length of the rotameter tube is 300 mm and its diameter at the inlet is 25 mm. Calculate the taper angle of the tube for drag coefficients of 0.3, 0.7, 0.9 and 1.20.

6. Water flows in a tube of 25 mm diameter at 3 m/sec. Assuming a one dimensional frictionless flow calculate the velocity measured by a 12 mm diameter Pitot static tube. What diameter Pitot static tube should be used to reduce this error to 1%?

7. Water is flowing through a 50 mm diameter pipe. A static pressure tube flush with the side wall of the flow pipe shows a pressure of 250 mm of water. The Pitot tube at the axis shows a pressure of 1000 mm of water. Calculate the flow rate.

8. A hot wire anemometer with time constant τ working in constant-current mode is to be used to measure velocity fluctuations upto 1 MHz. Discuss the compensation net work to achieve this.

9. A divergent beam of light is used for shadowgraphy. This object is at a distance L_0 from the point source and the screen is placed at a distance of L_1 from the object. Obtain an expression for the geometric sensitivity for shadow method.

10. The sensitivity of a Schlieren system is defined as the fractional deflection at the plane P_f per unit deflection of a light ray at the test section. Obtain an expression for the constrast in terms of sensitivity, density gradient, and the width of the test section.

11. An interferometer is used for the visualisation of a free convection boundary layer. For this application, the following data were collected:

$$\text{Plate temperature} \qquad T = 50°C$$

$$\text{Free stream air temperature} \qquad T = 20°C$$

$$\beta = 0.000293$$

$$L = 450 \text{ mm}$$

$$\lambda = 546 \text{ nm}$$

Reference density is taken at 20°C and 1.013 N/m² pressure. Calculate the number of fringes which will be viewed in the boundary layer. Calculate the temperature corresponding to the four fringes nearest to the plate surface.

12. Show that the sensitivity of an interferometer, defined as the number of fringe shifts per unit change of density, may be written as

$$S = \frac{\beta L}{\lambda \rho_s}$$

Show that the maximum sensitivity of an interferometer defined as the number of fringe shifts per unit change in Mach number will be about 30 when $L = 150$mm, $\lambda = 540$ nm, $\beta = 0.000293$ and the stagnation density is that of air at standard conditions.

13. A spherical ball is suspended by a fine string from a fixed point, Fig. 7.25. The wind moving at velocity V causes the string to be at an angle ϕ relative to vertical. Determine a relation between ϕ (degrees) and V (km/h).

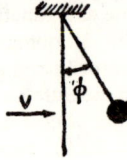

Fig. 7.25

14. A liquid flow meter consists of a circular bend of radius R in a pipe. Static pressure P_1 and P_2 are measured in the straight section and in the curved section of the pipe respectively.

Determine the relationship $Q = f(\rho, R, P_1, P_2)$ where ρ is the density of the liquid. What are the probable sources of error?

Measurement of Temperature

8.1 Introduction

The concept of temperature is a very difficult one. Temperature is usually defined as a measure of 'degree of heat' while heat is taken to mean the quantity of heat. It is equivalent to potential in electricity and level in hydrostatics. Higher temperature would mean that the heat would flow from it to the lower temperature irrespective of the heat content of the body at the higher temperature. The temperature can, therefore, be said to be 'heat level'.

On the basis of kinetic theory, the temperature of a system may be defined through the equation

$$\tfrac{1}{2}mv_{av}^2 = kT$$

where m is mass of the molecule, v_{av} its average velocity and k is the Boltzmann's constant. The equation holds good for systems which obey Maxwell Boltzmann distribution, i.e., for gases. However, for liquid and solid systems, the temperature may again be defined by the intensity of molecular activity but the relation is very complicated due the presence of strong inter-molecular forces. Therefore, temperature may be considered as a manifestation of the molecular kinetic energy of the body. All motions stop at absolute zero of temperature. There are, however, other definitions of the temperature also which have been proposed by various researchers. Two of them are:

 (i) A condition of a body by virtue of which heat is transferred to or from other bodies.

 (ii) A quantity whose difference is proportional to the work from a Carnot engine operating between a hot source and a cold receiver.

Since temperature is a derived quantity no such standard as the standards of mass, length, etc., can be defined. The temperature is thus measured by the measurement of certain properties of matter which are influenced by the degree of heat. The most used are changes in:

 (i) physical state,
 (ii) chemical state,
 (iii) dimensions,

(iv)   electrical properties,

(v)   radiation properties.

None of these changes gives an absolute measure of temperature but must be calibrated to some arbitrarily chosen standard.

8.2   Expression of temperature

The temperature is expressed by degree Celsius, Fahrenheit or Rankine. All these scales depend on the properties of the substance. In fact, on Celsius scale the difference between the freezing point and boiling point of water at NTP has been divided into 100 equal parts, each is called as one degree Celsius. The absolute temperature scale proposed by Lord Kelvin does not depend on the properties of the substance used and makes the basis of all thermodynamical quantities.

The international temperature scale of 1948 sets the basis of an experimental scale and has some well defined standard points as close as possible to the thermodynamic scale. The following six basic fixed points were adopted:

(i)   Oxygen point-boiling point of liquid oxygen ($-182.97°C$)

(ii)   Ice point-melting point of pure ice (fundamental point) ($0°C$),

(iii)   Steam point-boiling point of pure water (fundamental point) ($100°C$),

(iv)   Sulphur point-boiling point of liquid sulphur ($444.60°C$),

(v)   Silver point-melting point of silver ($960.8°C$)

(vi)   Gold point-melting point of gold ($1063.0°C$).

All these definitions of fixed points are at one standard atmosphere and at an equilibrium of the two phases. These points are used to standardise thermometers in the laboratory.

The points on the scale between the fixed points are interpolated as follows:

1.   From $-190°C$ to $0°C$: The temperature in the range $-190°C$ to $0°C$ is obtained by the measurement of resistance of a standard platinum resistance thermometer. The resistance R_t, at any temperature, t, within this range is given by

$$R_t = R_0 \left[1 + At + Bt^2 + C(t-100)t^3\right]$$

where the constants R_0, A, B and C are obtained by measuring the resistance at oxygen, ice, steam and sulphur points.

2.   From $0°C$ to $630.5°C$: The range between $0°C$ to $630.5°C$ is also covered by the measurement of resistance of a standard platinum resistance thermometer. However the resistance, R_t, at any temperature, t, within this range is given by

$$R_t = R_0 \left[1 + At + Bt^2\right]$$

where the constants R_0, A and B are obtained by measuring resistance at ice, steam and sulphur points.

3. From 630.5°C to 1063.0°C: The temperature in the range between 630.5°C and 1063.0°C is obtained by the measurement of emf, e, of a standard platinum, platinum-rhodium thermocouple, one junction being kept at the ice point, while the other junction at the temperature t °C in this range. The emf, e, is given by

$$e = a + bt + ct^2$$

where the constants a, b and c are obtained by measuring emf's at the freezing point of antimony (630.5°C), silver point and gold point.

4. Above 1063.0°C: The temperature of a body above the gold point is measured by comparing the intensity of radiation of a particular wave length emitted by the body, with the radiation emitted by the black body at the gold point. For temperatures exceeding 6300 K, the intensity of radiation is compared at microwave frequencies.

For temperatures below the oxygen point, the international practical temperature scale is not defined. However, the National Bureau of Standards has standardised thermometers upto 2 K, although they are not internationally accepted. They include the acoustical interferometer, the helium-vapour thermometer, the platinum resistance thermometer etc. The ^{3}He vapour scale of 1962 also known as T_{62} (^{3}He) scale covers the range from 0.8 K to 1.5 K and is based on the relation between pressure of saturated ^{3}He vapour on temperature. The ^{4}He vapour scale of 1958, T_{58} (^{4}He), spans the temperatures from 1.5 K to 4.2 K and is based on the relation between the pressure of saturated ^{4}He vapour on temperature.

Apart from the primary standard points described above, there are quite a few secondary points in the International scale of 1948. Some of them are given below:

 (i) Temperature at which solid carbon dioxide changes to gaseous CO_2 at normal atmospheric pressure (-78.50°C),
 (ii) Freezing point of mercury (-38.87°C),
 (iii) Freezing point of antimony ($+630.5$°C),
 (iv) Freezing point of palladium ($+1552$°C),
 (v) Melting point of tungsten ($+3380$°C).

The approximate range and accuracy of various temperature measuring devices have been given in Table 8.1.

The instruments for measuring temperature have been classified, in the first place, according to the nature of change produced in the testing body by the change of temperature. The following four broad categories have, therefore, been proposed:

 (i) Expansion thermometers
 (ii) Change of state thermometers
 (iii) Electrical methods of measuring temperature
 (iv) Radiation and optical pyrometry.

TABLE 8.1

No.	Type	Range, °C	Accuracy, °C
1.	Glass thermometers		
	(i) mercury filled	−39 to 400	0.3 to 1
	(ii) (Hg + N₂) filled	−39 to 540	0.3 to 5.5
	(iii) alcohol filled	−70 to 65	0.5 to 1
2.	Pressure-gauge thermometers		
	(i) vapour pressure type	11 to 200	1 to 5.5
	(ii) liquid or gas-filled	−150 to 600	1 to 5.5
3.	Bimetallic thermometers	− 74 to 549	0.3 to 14
4.	Resistance thermometer	−240 to 980	0.003 to 3
5.	Thermistors	−100 to 260	Depends on ageing
6.	Thermocouples		
	(i) base metals	−185 to 1,150	0.3 to 11
	(ii) precious metals	−185 to 1,150	0.3 to 11
7.	Pyrometers		
	(i) optical	760 and above	11 for black body
	(ii) radiation	540 and above	11-16 for black body
8.	Fusion	590 to 3600	As low as 20-30 under optimum conditions

8.3 Expansion thermometers

The thermometers falling under this category have been further subdivided as follows:
1. Expansion of solids
   (a) Solid-rod thermostats/Thermometers
   (b) Bimetallic thermostats/thermometers
2. Expansion of liquids
   (a) Liquid-in-glass thermometers
   (b) Liquid-in-metal thermometers
3. Expansion of gases
   (a) Gas thermometers

EXPANSION OF SOLIDS

Solid-rod thermostats/thermometers

Solid thermostat/thermometer is based on the differential expansion of materials i.e. a material will expand more than the other when exposed to the same change in temperature. One of the thermostats consists of a combination of an invar rod inside a brass tube: one end of the rod and tube is hard soldered. The other end carries a micro switch, and a micrometer etc. to initially adjust the gap (hence pre-set the temperature) between the free ends of the rod and the tube. When this combination is placed in a temperature bath, the brass tube ($a = 34.2 \times 10^{-6}$cm/cm°C) will expand more than the invar rod ($a = 2.7 \times 10^{-6}$cm/cm°C) so that the position of the free end of the rod changes with respect to the position of the free end of the tube. This differential change may actuate a micro switch to cut off the electric supply when the pre-set temperature is reached. To pre-set the temperature the micro-switch position relative to free end of the invar rod is adjusted by a micrometer. This kind of thermostat is used to cut off the supply to electric heaters ond ovens. It can also be used to monitor the supply of gas to gas ovens.

Bimetallic thermostats/thermometers

A bimetallic thermometer consists of two metal strips having different coefficients of expansion. These strips are brazed together at a temperature t_b called the brazing temperature. A change in temperature causes a free deflection of the assembly as shown in Fig. 8.1. When the temperature to be measured is higher than the brazing temperature, it bends towards the low coefficient side, while if the temperature is less than the bonding temperature, it bends in the other direction. Such bimetallic strips are the basic elements for many control devices and are

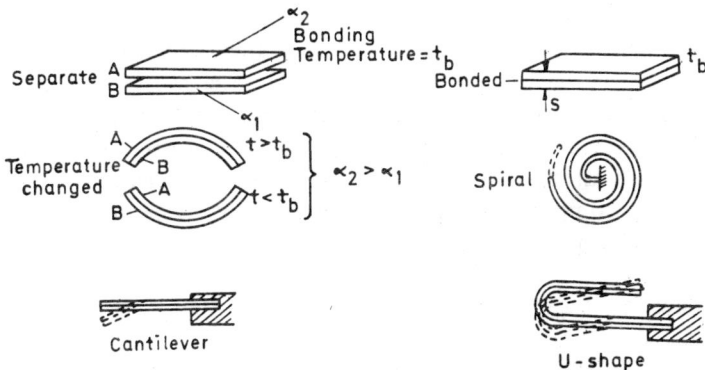

Fig. 8.1 Some bimetalic sensors

also used to some extent for the temperature measurement.   The follow-ing relation may be used to determine the radius of curvature of such a strip which is initially flat at the temperature t_b,

$$r = \frac{s\left[3(1+m)^2 + (1+mn)\left(m^2 + \frac{1}{mn}\right)\right]}{6(a_2 - a_1)(t - t_b)(1 + m)^2}$$

where s = combined thickness of the bonded strip,

m = ratio of the thickness of low expansion to that of the high expansion material,

n = ratio of modulus of elasticity of low expansion to that of the high expansion material,

a_1 = linear expansion coefficient of low expansion material

a_2 = linear expansion coefficient of high expansion material, and

t = temperature of the medium.

In most practical cases $m \simeq 1$

and $n + (1/n) \approx 2$

Thus the above equation reduces to

$$r \simeq \frac{2s}{3(a_2 - a_1)(t - t_b)}$$

This expression can be used to calculate the deflections of various kinds of bimetallic elements.   The design is such that the deflection is propor-tional to the temperature difference.   One of the materials used for bimetallic strips is generally Invar.   The low expansion strip is made of it while the high expansion strip is made of brass or a variety of other alloys.

Thermometers with bimetallic sensors are often used for the measure-ment and control of temperature because of their ruggedness and some-times because of their convenient shape.   These elements can be fabri-cated in the cantilever form, U form, helical form, spiral and washer forms.   Whenever greater accuracy is desired bimetal strips in the helical or spiral form are used.   These elements are also used for temperature compensation in some instruments where temperature is either a modifying or an interfering input.

EXPANSION OF LIQUIDS

Liquid-in-glass Thermometer

The liquid-in-glass thermometer is one of the most common types of temperature measurement devices.   It consists of a very big bulb   with very thin walls to hold the temperature sensing liquid. The bulb is connected to a uniform bore capillary which is graduated. On the other end of the capillary is another small bulb called a safety bulb. Figure 8.2 shows a sketch of a thermometer. Mercury and alcohol are most commonly used liquids. The thermometer works on the principle of differential expansion of liquid. The following points may be borne in mind while designing a thermometer:

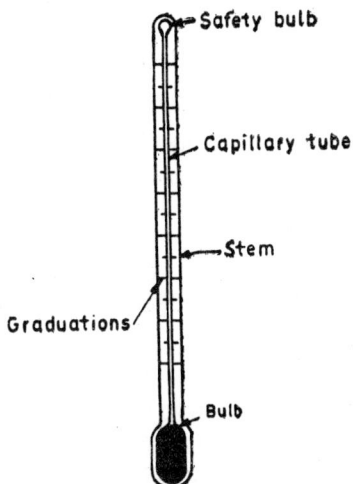

Fig. 8.2 Mercury-in-glass thermometer

(i) Walls of liquid bulb should be thin, thereby quick transfer of heat is possible and hence the fast response provided the volume of the liquid is small. For higher sensitivity, the volume of liquid should be large, thus the response is impaired. A compromise is, therefore, to be made between sensitivity and response.

(ii) Capacity of bulb is many times the volume of capillary and hence a safety bulb at the other end is provided.

(iii) Alcohol may be preferred over mercury due to its higher coefficient of expansion.

Mercury-in-glass thermometers are generally employed up to 340°C (boiling point of mercury is 357°C) but their range may be extended to 560°C by filling the space above mercury with CO_2 or N_2 at high pressure, thereby increasing its boiling point and hence the range. The accuracy of these thermometers under optimum conditions does not exceed 0.1°C. However, when an increased accuracy is required, Beckmann thermometer can be used. It contains a big bulb attached to a very fine capillary. The range of the thermometer is limited to only 5 to 6°C with an accuracy of 0.005°C

The thermometers are used in two ways-complete immersion, and partial immersion. The complete immersion thermometers are calibrated to read correctly when the liquid column is completely immersed in the temperature medium.   Since this obscures the reading of the scale, a little portion of the liquid column is allowed to project a little to permit observation.

The partial immersion thermometers are immersed to a definite liquid column and the exposed portion is to be at a temperature for which calibration is valid.   However, if the temperature of the stem is different from that used during calibration, proper correction can be applied as follows:

$$\text{correction} = 0.00016(t_1 - t_2)n \text{ °C scale,}$$

$$= 0.00009(t_1 - t_2)n \text{ °F scale,}$$

where t_1 is the temperature of the stem, t_2 is the calibration temperature and n is the number of degree exposed.   The numerical value is the coefficient of apparant expansion of mercury in glass.

Liquid-in-metal Thermometers

Mercury-in-Steel Thermometer: Two distinct disadvantages of liquid-in-glass thermometers are:

(i)   glass is very fragile and hence immense care should be exercised in handling these thermometers, and

(ii)   the position of thermometer for accurate temperature measurement is not always the best position for reading the scale of the thermometer.

Both of these disadvantages are overcome in mercury-in-steel thermometer.   The principle of operation is again differential expansion of liquid which is used in a modified form as explained below:

Since metals are opaque, i.e., mercury column is not visible, the change in volume due to temperature change is thus obtained via the pressure change which is read off by the Bourdon tube (Fig. 8.3).   The thermometer bulb, capillary and Bourdon tube are filled with mercury usually at high pressure.   When the thermometer is used for the measurement of temperature change, the volume of mercury increases, which

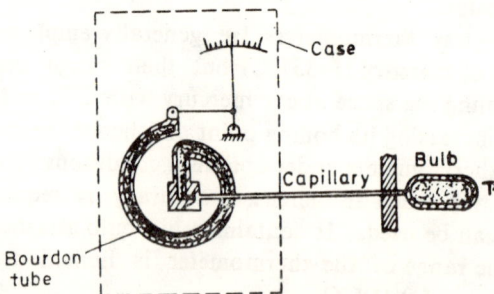

Fig. 8.3   Pressure thermometer

exerts a force (pressure) on the Bourdon tube. Through a rack and pinion arrangement, the deformation (displacement) of the Bourdon tube is magnified and read on a scale. The scale is graduated in degrees.

Various forms of Bourdon tubes, say of helical or spiral form, are used to increase the sensitivity. For easy and safe operation, a very long pressure transmitting element, i.e. a capillary is used between the bulb and the Bourdon tube. Since capillary is very long, the effect of ambient temperature may be very serious and may require compensation. Figure 8.4 shows two temperature compensated arrangements: in one spiral Bourdon tube and in other C-type Bourdon tubes are used. The length of the capillary in the compensating Bourdon tube is equal to that of the measuring thermometer and runs parallel to it. Ambient temperature variations influence both the capillaries equally and hence any pressure thus developed imparts the displacement to Bourdon tubes in opposing sense. Another method is offered by Taylor Instruments Co. for ambient temperature compensation, which is called the Accuratus mercury actuated tube system. The system contains a filler wire in the stainless steel Accuratus tubing. The filler wire extends throughout the bore of the tubing. This wire is made of invar metal (low temperature coefficient). The diameter of the wire is so chosen that it decreases the volume of mercury in the capillary to the extent where the volumetric expansion of mercury with a given increase in the ambient temperature exactly equals the increase in volume of the metal capillary due to the same ambient temperature increase. Thus no pressure is developed in the Bourdon tube due to ambient temperature changes.

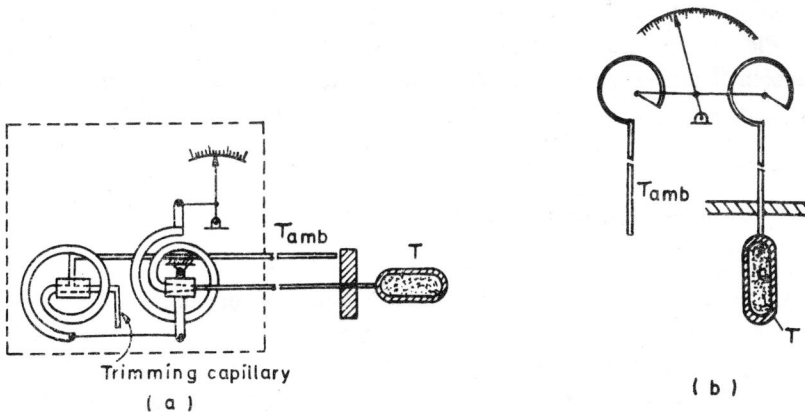

Fig. 8.4   Ambient temperature compensation methods

EXPANSION OF GASES

Gas thermometer

It is well known that the volume of a gas increases with temperature if the pressure is maintained constant, and the pressure increases if the volume is maintained constant. In a constant volume thermometer, the mass and the volume of the gas are kept constant, and hence the pressure of the gas changes as the temperature changes. On the other hand, in constant pressure thermometer, the mass and the pressure of the gas are kept constant, and hence the volume of the gas changes as temperature changes. At constant volume, the pressure of a gas as a function of temperature is given by

$$p_T = p_0 (1 + \beta T),$$

where p_T and p_0 are the pressures at temperatures $T°C$ and $0°C$ respectively and β is the thermal coefficient of pressure. The pressure change Δp in the gas pressure due to the change in temperature from T_1 to T_2 is given by

$$\Delta p = p_0 \beta(T_2 - T_1).$$

The pressure change is directly related to the temperature change. This holds good for ideal gases. In practice nitrogen is used as a filling gas, and hence corrections for departure from ideal behaviour should be supplied.

A gas thermometer, therefore, contains a certain volume of gas enclosed in a bulb, a capillary and a Bourdon tube. The pressure indicated by the Bourdon tube is calibrated in terms of the temperature of the bulb. The change in pressure per degree change in temperature is very small. This limits the actuating power of the Bourdon tube. The Bourdon tube is therefore used only to move the pen rather than for temperature compensation where compensating Bourdon tube is to be driven. Thus these thermometers are compensated only for case. Further the bulbs of these thermometers are made large so that the volume of gas in the bulb is very large compared to that of capillary, thus reducing the effect of ambient temperature. But this reduces the dynamic response of the thermometer for transient changes. Still its response is faster compared to liquid thermometers of the similar bulb size.

8.4 Change-of-state thermometers

LIQUID IN METAL THERMOMETERS

Consider a container having a certain quantity of liquid, the space above it is assumed to be empty. The molecules of liquid are always in the state of random motion, moving in all the directions with different velocities. If a molecule has a vertical component of kinetic energy greater than the force of attraction at the liquid surface, it escapes from the liquid. Thus there would be a stream of liquid molecules escaping. This constitutes vapour. On the other hand, the molecules which constitute vapour are also moving at random and a few are attracted by the liquid. Thus the processes of evaporation and condensation go on simultaneously. When the rates of evaporation and condensation are equal, the vapour becomes saturated.

It is to be noted now that the vapour is always saturated when in contact with the liquid. If the size of the upper part of the container is altered, the vapour pressure will fall momentarily with an increase in volume and rate of evaporation would increase till the vapour is saturated and vice versa. Thus the saturated vapour pressure depends only on the temperature and properties of the liquid and is independent of the size of the container. A typical graph illustrating the variation of saturated vapour pressure of water with temperature is shown in Fig. 8.5. The relationship is non-linear.

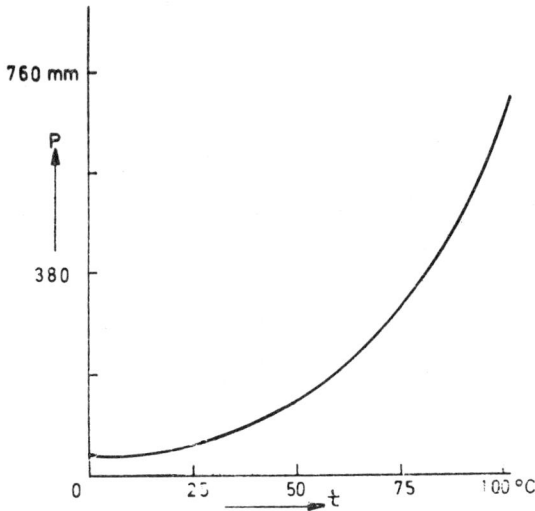

Fig. 8.5 Variation of vapour pressure of water with temperature

The vapour pressure thermometer consists of a bulb containing the fill liquid, a capillary tube and the Bourdon tube. Various kinds of liquids like propane, sulphur dioxide, ethyl ether, methyl chloride, and toluene are used as fill liquids depending on the range of the instrument. If the temperature of the bulb is higher than that of the instrument (i. e. capillary and Bourdon tube), the capillary and a part of the Bourdon tube may be filled with the liquid. On the other hand, when the bulb temperature is smaller than that of the instrument, vapour liquid interface lies in the bulb itself. The accuracy of the vapour-pressure thermometer is not influenced by the ambient temperature variations provided these do not oscillate around the process temperature. Suppose the process temperature is between 15.5°C and 18.5°C, and the ambient temperature around the capillary and the Bourdon tube fluctuates between 10°C and 25°C. Further assume that the ambient temperature is 10°C, and it suddenly changes to 25°C. This will cause an increase in the vapour pressure and hence momentairly upscale movement of the pointer indicating higher temperature although process temperature has not changed. However, the condensation of vapour will take place till the equilibrium is reached and the pointer will move downscale to indicate the process temperature. Similarly, a different temperature is indicated momentarily when the ambient temperature swings from high to low temperature side about the process temperature. Besides there is an error introduced by the head effect due to cross-ambient temperature.

In order to circumvent these problems, a non-volatile liquid is used in addition to the volatile liquid.   Figure 8.6 shows one such thermometer.

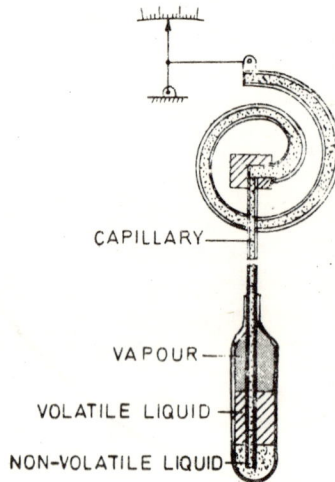

CAPILLARY

VAPOUR

VOLATILE LIQUID

NON-VOLATILE LIQUID

Fig. 8.6   A vapour pressure thermometer

The non-volatile liquid fills the Bourdon tube and the capillary and the volatile liquid with its vapour always remains in the bulb. As the temperature of the bulb increases, the vapour pressure of the volatile liquid increases. This pressure is transmitted to the Bourdon tube through the non-volatile liquid. The influence of ambient temperature is extremely small as the non-volatile liquid remains in liquid phase even at the high ambient temperatures. The relationship between vapour pressure and temperature is highly non-linear, and is univalued only below a certain temperature called the critical temperature. Therefore, the range of a vapour pressure thermometer is restricted by the critical temperature of the fill liquid. The range can be extended by using a variety of other liquids. Further the scale can be linearised by employing special linearising linkages. The choice of materials for fabricating the bulb is also very wide: metals and their alloys may be used. These may be plated with nickel, chromium or silver if necessary. The vapour-pressure thermometers are probably the most widely used thermal systems being the fastest in response and lowest in cost. The advantages and limitations of these systems are as follows:

(i) They are cheaper than mercury-in-steel thermometers,
(ii) Their dynamic response is better than those of liquid-in-metal thermometers,
(iii) They do not suffer from ambient temperature effects, and
(iv) They can be constructed with smaller bulbs than those used in other types of thermometers.

Limitations

(i) They have a limited linear range due to non-linear response,
(ii) Further their range is limited by the fact that the minimum temperature for which they can be used must be well below the critical temperature.

PYROMETRIC CONES

A material changes its physical state on heating. The temperature at which this change occurs depends on the purity of the material and the pressure. There are a number of temperature sensing devices which are based on change of state of the material. The melting points of certain minerals and their mixtures are used to check the temperature of a kiln in the ceramic industry. Cones known as Seger cones are made of these minerals in different compositions so that they melt at different temperatures. By varying the composition of the cones, a range of temperature from 600°C to 2000°C can be covered in steps of 25°C to 45°C. To check the temperature, a series of cones of different compositions is placed in the kiln. Cones having a lesser melting temperature will melt, eventually a cone will be found which just bends over. This cone indicates the temperature of the kiln. The maximum sensitivity is $\pm 10°C$.

Somewhat similar temperature indicators are also available which indicate the temperature by a change of colour or appearance. These are available in the form of lacquers, sticks and pellets. They cover a range from 45°C to 1370°C in steps of 7°C to 25°C. Those based on change of colour exhibit a particular colour when a certain temperature is reached, while those based on change of appearance change from dull and chalky state to shiny and liquid state at a predetermined temperature. These temperature sensors are used to indicate the temperature of electrical equipment, retorts and piping etc.

8.5 Electrical methods

In electrical methods of measuring temperature, the temperature signal is converated into electrical signal either through a change in resistance (or voltage), leading to a change in current or development of emf. The following elements are used to convert temperature into electrical variables:

1. Electrical resistance bulbs,
2. Thermistors, and
3. Thermocouples and thermopiles.

ELECTRICAL RESISTANCE BULBS/THERMOMETERS

Most of the work in connection with electrical resistance thermometry involves the knowledge of Ohm's law. It is found that the resistance of pure metallic conductors increases with temperature. In practice, platinum, copper and nickel are used as resistance elements because they can be obtained in a high degree of reproducibilities of resistance characteristics. These elements can be used in the following ranges:

Platinum — 190°C to + 630.5°C,

Copper — 50°C to + 250°C,

Nickel — 200°C to + 350°C.

At temperatures approaching absolute zero, leaded phosphor bronze resistance thermometers are found to be more suitable. Thermometers using platinum as the resistance element are discussed here.

Platinum Resistance Thermometer

Platinum is a standard material for the resistance element in the resistance thermometer that defines the International Temperature Scale of 1948. It has certain desirable features like it is stable, it can be drawn into fine wires, it can be obtained in an extremely high purity and has a very high melting point. On the other hand, resistance-temperature relationship of platinum can be altered considerably by contaminations and mechanical strains. In other words, temperature coefficient of resistance is sensitive to impurity and strains. It is therefore essential to anneal the resistance element at temperature higher than the working temperature to release internal strains. The combination of purity and adequate annealing is shown by the ratio of resistances of the element at the steam and ice points. The ratio of 1.3910 of higher is the accepted value according to the International Temperature Scale of 1948.

Platinum Scale

The resistance-temperature relationship of the platinum thermometer is non-linear. If the resistance-temperature relationship is regarded linear, the temperature may be expressed in terms of resistance change. This constitutes the platinum scale of temperature. This scale approximates the thermodynamic scale.

The resistance R_t of the platinum resistance element at any temperature t on platinum scale is expressed by the equation

$$R_t = R_0 (1 + \alpha t),$$

where the temperature coefficient of resistance α, assumed constant on platinum scale, is obtained by measuring resistances R_{100} and R_0 at steam (100°C) and ice (0°C) points respectively. That is

$$\alpha = \frac{R_{100} - R_0}{100 \, R_0}$$

The difference $(R_{100} - R_0)$ between the resistance values at 100°C and 0°C is called the 'fundamental internal' of the thermometer. The temperature t_p as measured on the platinum scale may be determined by measuring the resistance R_t at the required temperature from the equation

$$t_p = \frac{R_t - R_0}{R_{100} - R_0} \times 100.$$

The temperature obtained on platinum scale is obtained under the assumption that the resistance-temperature relationship is linear. It has, however, been pointed out earlier that the resistance of a platinum thermometer, in the temperature range of 0°C to 630.5°C on International Temperature Scale, is non-linear and is given by

$$R_t = R_0 [1 + At + Bt^2]$$

Therefore, the temperature t_p on platinum scale for the measurement of temperature above 100°C will be less than that given on the International Temperature Scale. Thus, a correction is to be applied to the platinum scale. The correction term is obtained as follows:

It has been assumed that the temperature coefficient of resistance is constant upto 100°C. Under this assumption both the equations should give identical results upto 100°C. Therefore at 100°C,

$$R_0 (1 + 100\alpha) = R_0 (1 + 100 A + 100^2 B)$$

This gives

$$\alpha = (A + 100 B).$$

For temperature above 100°C, the two scales give different results. However, the thermometer is exposed to the same temperature which are expressed as t_p and t_I on the two scales, and thus has the same resistance R_t. Therefore

$$R_t = R_0 (1 + At_I + Bt_I^2) = R_0 (1 + \alpha t_p).$$

The correction term $(t_I - t_p)$ can now be written as

$$t_I - t_p = \delta [(t_I/100)^2 - (t_I/100)], \text{ where}$$

$$\delta = -\frac{100^2 B}{A + 100 B}.$$

The temperature on the International scale is, therefore, obtained by adding the correction term to the platinum temperature as,

$$t_I = t_p + (t_I - t_p).$$

This is known as Callender formula and holds good over a large temperature range above 100°C. The temperature on international scale from platinum scale is obtained by the method of successive approximation. The value of constant δ depends on the purity of resistance element, and for pure platinum its value is 1.494. A similar equation which holds good between 0°C and −200°C can also be derived. It is known that the resistance, in this range, is given by

$$R_t = R_0 \left[1 + At_I + Bt_I^2 + C (t_I - 100)^3 \right]$$

The difference between the temperatures on the international and platinum scales can be shown to be

$$t_I - t_p = \delta \left[\frac{t_I}{100} - 1 \right] \frac{t_I}{100} + \beta \left[\frac{t_I}{100} - 1 \right] \frac{t_I}{100} ,$$

where β is another constant whose value depends on A and B. This equation is known as Callender-Van Dusen equation and is used to obtain temperature on international scale.

Resistance Thermometer Bulb

Resistance thermometer bulbs are either tip-sensitive or stem-sensitive. The tip-sensitive thermometers are used for the measurement of surface temperatures while stem-sensitive are used for other applications.

In modern stem-sensitive thermometers, the coil consists of pure platinum wire wound on a former of mica, stealite, or porcelain so that it is completely free from strain. Figure 8.7 is a cut-section of the stem of a thermometer. The resistance wire is wound on mica former which is supported from both sides by springs. The spring provides a resilient cushion against vibration. It also provides good-thermal contact between the resistance element and the shell of the thermometer, For a tip-sensitive resistance thermometer the platinum wire is wound again on a former, say of mica, which is wrapped round a solid-silver core. The heat is conducted rapidly from the end flange to the resistance element through the core.

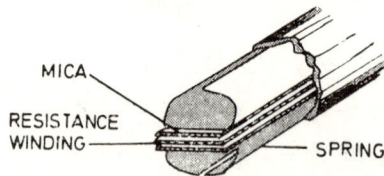

Fig. 8.7   Resistance thermometer bulb

After winding, the resistance element is annealed, to relieve internal strains, at a temperature higher than the maximum temperature encountered in service. Elements of fundamental interval of 1Ω or 10Ω are commonly used. These elements are made of fine wire of platinum of diameters ranging from 50 μm to 100 μm for the temperature range of −200°C to 750°C. If the thermometer is to be used at higher temperatures around 1000°C, a platinum wire of 500 μm diameter is usually used. The ends of resistance element are connected to heavy copper or constantan leads by means of gold wires to reduce the influence of thermo emf due to joining of two dissimilar metals. For resistance elements having higher resistance the influence of the resistance of leads and thermo emf is negligibly small. The leads are insulated from each other and from the shell by means of beads or discs of porcelain. However all precautions should be taken to avoid contact resistance, and thermo emf by keeping the junctions at the same temperature.

Measurement of Resistance

To measure the temperature, resistance of the coil is measured by some form of the bridge. The bridge can be used either in the null or deflection position. Null method, however, gives better results. Since sensing element is some distance away, copper or constantan cables are used to make connections. The leads should have resistance very much smaller compared to the resistance of the element. Further, any variation of resistance due to ambient temperature should be compensated for. The following methods are often used:

Three-Lead Arrangement: This was introduced by Siemens in 1871 and is shown in Fig. 8.8 (a). Three leads l_1, l_2, l_3 are connected to the resistance element and the bridge circuitry. The lead l_1 is in the same arm of the bridge as the resistance R while l_3 is in the thermometer arm. The lead l_2 connects the bridge point D to one terminal of the element. It will be seen that if l_1 and l_3 are identical wires and have the same length and cross section, they will have equal resistances, and any change of ambient temperature will affect them equally, thus the bridge always remains balanced. Further, when the bridge is balanced, no current flows through the lead l_2. The effect of the variable and unknown contact resistance in the adjustable resistor R has no influence on the resistance of the bridge arm at null balance.

Callender's Method: The arrangement of connections is shown in Fig. 8.8(b). Two identical pairs of leads are connected in arms AD and

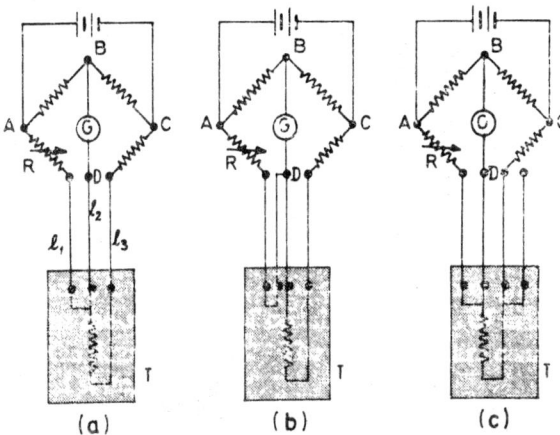

Fig. 8.8 Three methods for compensating lead resistance

CD of the bridge. The pair in the arm AD is connected together at a point near the thermometer bulb, while the second pair which is in the arm CD, is connected to the thermometer bulb. Both pairs of leads are enclosed in the same outer cover so that the leads in AD compensate for any changes in the resistance of the leads in CD due to ambient temperature variations. This method is quite useful when thermometers are used in both arms to measure differential temperature.

Four-Lead Arrangement: This arrangement is shown in Fig. 8.8(c) and is used in the same way as the one with three leads. Provision is made, however, for using any combination of three, thereby checking for unequal lead resistances. By averaging the readings, more accurate results are possible. Some form of this arrangement is used where highest accuracy is required.

The principle of the Wheatstone bridge allows the replacement of resistances by inductances and capacitances. In one variant of resistance thermometer, two arms of the bridge are formed by two capacitors, one of them being the balancing capacitor. This reduces the influence of contact resistance.

The thermometers are excited by a d.c. or a.c. source of voltage; the current flowing in the resistance element is usually in the range of 2 to 20 mA. This current causes v^2R heating which raises the temperature of the thermometer bulb, above that of the surroundings, causing the so-called self-heating error. The magnitude of this error is usually very small, as an example a 450-ohm element in open construction carrying 26 mA current has a self-heating of 0.1°C when immersed in liquid oxygen. The error can be reduced by pulse excitation of the bridge.

THERMISTORS

Thermistor is a temperature sensitive variable resistor made of a ceramic like semi-conducting material. They are available in a resistance range from ohms to mega ohms and their resistive characteristics, coupled with

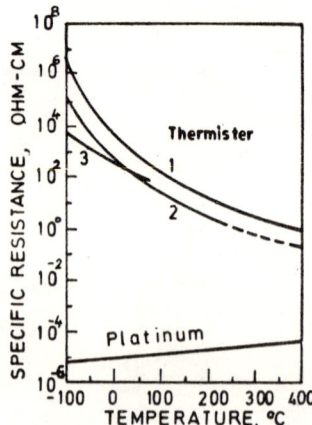

Fig. 8.9   Resistance-temperature characteristics of thermistors

stability and high sensitivity, make them a highly versatile tool for temperature measurement. Unlike metals, thermistors respond negatively to temperature, and their temperature coefficient of resistance is about 10 times higher than that of platinum or copper. Figure 8.9 shows resistance-temperature characteristics of thermistors along with that of platinum.

Thermistors are made of metal oxides and their mixtures, namely oxides of copper, nickel, manganese, iron, tin, cobalt etc. The oxides and their mixtures in powder form are compressed into desired forms and sizes, and than heat treated for making the thermistors. The electrical contacts can be made in a number of ways, namely (i) lead wires are embedded in the material before heat treatment, and (ii) wires are attached to the plating or metal ceramic coating on the thermistor. They are available in various forms like beads as small as 0.4 mm in diameter, discs ranging from 5 to 25 mm in diameter, and rods from 0.8 to 6.0 mm in diameter and upto 50 mm in length (Fig. 8.10). Flakes (only a few micrometers thick) are employed as infrared radiation detectors or bolometers. Size and shape of a thermistor is decided by several factors, such as space available, the required speed of response to temperature changes and the amount of power dissipation. Various methods of mounting are used: beads are suspended from wire leads or imbedded in probes, discs are mounted on spring-loaded stacks with or without heat dissipating fins, other discs or rods are pigtail-mounted. Beads and small discs may be covered with a thin adherent coat of glass to reduce composition changes of thermistor at high temperature.

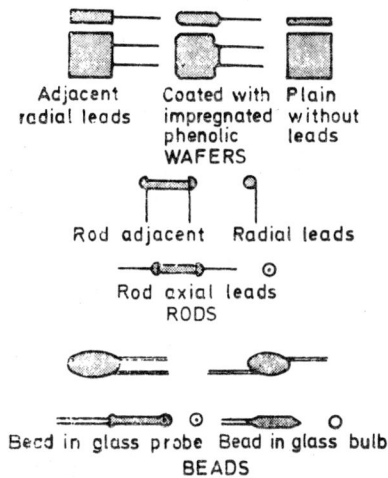

Adjacent
radial leads

Coated with
impregnated
phenolic
leads
WAFERS

Plain
without
leads

Rod adjacent   Radial leads

Rod axial leads
RODS

Bead in glass probe   Bead in glass bulb
BEADS

Fig. 8.10 Typical construction of thermistors

The temperature vs resistance relationship of the thermistor is given by

$$R_T = R_0 e^{\beta\left(\frac{1}{T}-\frac{1}{T_0}\right)}$$

where R_T = resistance at temperature TK,

R_0 = resistance at temperature $T_0 K$, and

β = constant.

The value of β usually lies between $3400\ K$ and $4600\ K$ depending on the composition. The temperature coefficient of resistance is

$$\frac{dR/dT}{R} = -\frac{\beta}{T^2}$$

Assuming $\beta = 4000\ K$ and $T = 298K$, $\left(\dfrac{dR}{dT}\right)\dfrac{1}{R} = -0.045/K$.

The value of $(dR/dT)/R$ for platinum is $0.0036/K$ indicating that the thermistor is at least 10 times more sensitive than the platinum resistance element. Further the stability of properly aged modern thermistors is quite acceptable for many applications.

The application of thermistors to the measurement of temperature follows the usual principles of resistance thermometry. Conventional bridge or other resistance measuring circuits are commonly employed. The high temperature coefficients of thermistors result in their having greater available sensitivity as temperature-sensing elements than the resistance thermometers or common thermocouples. The thermistors are used for temperature measurement, temperature compensation, temperature control and time delay applications.

THERMO-ELECTRIC THERMOMETRY

The thermoelectric thermometry is based on the fact that an emf is generated when the junctions of two dissimilar metals are kept at different temperatures; the magnitude of the emf depends on the temperatures of the junctions. Consider a circuit formed by joining conductors of two dissimilar metals A and B as shown in Fig. 8.11. If the temperatures of the junctions are T_1 and T_2, an emf E_s is generated, and if the circuit is closed a current flows. The magnitude of the emf depends on the temperatures T_1 and T_2 of the junctions, and the metals A and B. The overall relationship between the emf E_s and the temperatures T_1 and T_2 is called the Seebeck effect. For a combination of metals A and B, the Seebeck voltage dE_s for very small temperature difference dT is

$$dE_s = \alpha_{A,B}\, dT,$$

where $\alpha_{A,B}$ is the Seebeck coefficient.

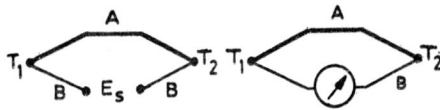

Fig. 8.11 Thermocouple circuit

The Seebeck emf generated in a thermo-electric circuit is due to two different effects, namely Peltier effect and Thomson effect. It was observed that a junction is cooled when the current is introduced externally in the circuit. If the direction of current is reversed, the junction is heated. The Peltier effect concerns the reversible evolution or absorption of heat at a junction between two dissimilar metals when an electric current crosses it. External heating or cooling of the junction generates an emf which is called the Peltier emf. The direction and magnitude of the Peltier emf depends upon the temperature of the junction and the metals forming the junction. It does not depend on the temperature of the other junction. It would now seem that both Seebeck effect and Peltier effect are the same phenomenon. It was however argued by Thomson that if an electric current produced only the reversible Peltier heating effects, then the net Peltier emf would be equal to the Seebeck emf, and would also be linearly proportional to the temperature difference at the junctions. This argument was at variance with the experimental observations that the relationship between emf and temperature is non-linear. Therefore, he concluded that Peltier emf is not the only source of thermo emf. A single conductor when exposed to a longitudinal temperature gradient must also develop an emf. The Thomson effect, therefore, concerns the reversible evolution or absorption of heat in a conductor having a temperature gradient when an electric current flows through it. The direction and magnitude of Thomson emf depends on the temperature, the temperature gradient and the material. This is, however, obvious that Thomson emf alone cannot sustain a current in a single homogeneous conductor forming a closed circuit, since equal and opposite emf's will be set up in the two paths from heated and to the cooled end. It may be noted that the magnitude of Thomson emf is considerably smaller than that of Peltier emf.

The total emf generated in a circuit is the result of four emf's, two due to Peltier effect (one at each junction) and two due to Thomson effect. It can be expressed as

$$E_s = \pi_{A,\,B/T_1} - \pi_{A,B/T_2} + \int_{T_1}^{T_2} \sigma_A dT - \int_{T_1}^{T_2} \sigma_B dT = \int_{T_1}^{T_2} a_{A,B}\, dT,$$

where $\pi_{A,B}$ is the Peltier emf and $\int \sigma dT$ is the Thomson emf. An arrangement of two dissimilar metals whose one junction is used for temperature measurement and other junction maintained at a reference temperature is called a thermocouple. The relation between the emf developed and the temperature is not linear and is different for different thermocouples. Each thermocouple is therefore to be calibrated. Of some interest is the slope of emf vs temperature curve or the thermo-electric sensitivity. It varies with temperature. The sensitivity is also called 'thermo-electric power' and is expressed $\mu V/°C$. Thermo-electric sensitivities of some materials relative to platinum are given in Table 8.2.

TABLE 8.2 Sensitivity of various materials relative to platinum
(Reference junction at 0°C)

Material	Sensitivity at 0°C ($\mu V/°C$)	Material	Sensitivity at 0°C ($\mu V/°C$)
Bismuth	− 72	Silver	+ 6.5
Constantan	− 35	Copper	+ 6.5
Nickel	− 15	Gold	+ 6.5
Potassium	− 9	Tungsten	+ 7.5
Sodium	− 2	Cadmium	+ 7.5
Platinum	0	Iron	+ 18.5
Mercury	+ 0.6	Nichrome	+ 25
Carbon	+ 3	Antimony	+ 47
Aluminium	+ 3.5	Germenium	+ 300
Lead	+ 4.0	Silicon	+ 400
Tentalum	+ 4.5	Tellurium	+ 500
Rhodium	+ 6.0	Selenium	+ 900

The emf can be measured either by a millivoltmeter or a potentiometer. When a millivoltmeter is used for the measurement of emf, a current flows in the circuit. According to Peltier law, this will result in the cooling of one junction and heating of the other. This will, therefore, lower the temperature of the hot junction. But the magnitude of this current is so low as to effectively change the temperature. However, a potentiometeric arrangement is normally preferred. Materials for thermocouples (commercial) are selected such that Thomson effect can be disregarded and the total emf is the sum of Peltier emf's only and thus depends only on the difference of the junction temperatures. If the temperature at one junction (reference junction) is kept constant, the emf generated is used to measure the temperature of the other junction or the temperature difference between the two junctions.

LAWS OF THERMOCOUPLES

The following laws are very useful as they govern both theory and practice of thermocouples:

(i) LAW OF HOMOGENEOUS CIRCUIT. *A thermo-electric current cannot be sustained in a circuit of a single homogeneous material, however varying in cross section, by the application of heat alone.* The consequence of this law is that two different materials are required for any thermocouple circuit.

(ii) LAW OF INTERMEDIATE MATERIALS. *Insertion of an intermediate metal into a thermocouple circuit will not effect the net emf, provided the two junctions introduced by the third metal are at identical temperatures.* This law suggests that a device for measuring thermo-electric emf may be introduced into circuit at any point without affecting the net emf, provided all the junctions which are added to the circuit by introducing the device are all at the same temperature. The proof of this statement is given in Fig. 8.12 (a).

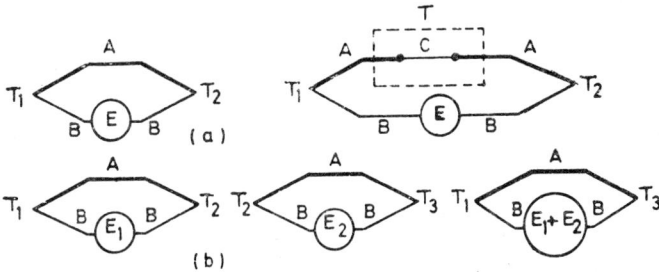

Fig. 8.12 Thermocouple laws

(iii) LAW OF INTERMEDIATE TEMPERATURES. *If a single thermocouple develops a net emf, E, when its junctions are at temperatures T_1 and T_2, and an emf, E', when its junctions are at T_2 and T_3, it will develop an emf $E_1 = E + E'$, when its junctions are at temperatures T_1 and T_3.* The proof of this statement is given in Fig. 8.12 (b). A consequence of this law permits a thermocouple calibrated for a given temperature to be used with any other reference temperature through the use of a suitable correction. Also, the extension wires having the same thermo-electric characteristics as those of the thermocouple wires can be introduced in the circuit without affecting the net emf of the thermocouple.

Thermo-electric Series

The various conductors have been tabulated in an order such that at a specified temperature each material in the list is thermo-electrically negative with respect to all above it and positive with respect to all below it.

TABLE 8.3   Thermo–electric series for selected metals and all alloys

100°C	500°C	900°C
Antimony	Chromel	Chromel
Chromel	Nichrome	Nichrome
Iron	Copper	Silver
Nichrome	Silver	Gold
Copper	Gold	Iron
Silver	Iron	$Pt_{90} Rh_{10}$
$Pt_{90} Rh_{10}$	$Pt_{90} Rh_{10}$	Platinum
Platinum	Platinum	Cobalt
Palladium	Cobalt	Alumel
Cobalt	Palladium	Nickel
Alumel	Alumel	Palladium
Nickel	Nickel	Constantan
Constantan	Constantan	
Copel	Copel	
Bismuth		

Note that at different temperatures, the sequence of materials changes. This is because in certain combinations and at certain temperatures called thermo-electric neutral points, Peltier effect is not apparent.

Thermocouple Materials

A thermocouple can be formed by any two dissimlar metals. Actually, of course, certain materials and combinations are better than the others. These combinations must possess reasonably linear temperature emf relationships, they must develop an emf per degree temperature change that is detectable with standard measuring equipment, they must be able to withstand high temperatures, rapid temperature changes and the effect of corrosive atmosphere. A few preferred materials are:

Copper Cu	Iridium Ir
Iron Fe	Constantan (60 Cu + 40 Ni) $Cu_{60}Ni_{40}$
Platinum Pt	Chromel (10 Cr + 90 Ni) $Cr_{10}Ni_{90}$
Rhodium Rh	Alumel (2Al + 94 Ni + 1 Si + 3 Mn)
	$Al_2Ni_{94} Si_1 Mn_3$

Performance characteristics of some of the thermocouples are given in Table 8.4.

TABLE 8.4   Characteristics of some thermocouples

	Thermocouple materials	Practical range	Inherent accuracy	Applications
1.	Pt-(Pt-10) Rh Pt-(Pt-13) Rh	0°C to 1450°C	0.1°C	Calibration standard where high accuracy at high temperature is required
2.	Cu-constantan	— 200°C to 350°C	0.2°C	Commercial oven temperature indication and control
3.	Chromel-alumel	— 200°C to 1100°C	0.5°C	Commercial furnace temperature indication and control
4.	Chromel-constantan	— 100°C to 1000°C	0.5°C	
5.	Iron-constantan	— 200°C to 750°C	0.5°C	

The thermocouples of noble metals like $Pt - (Pt - Rh)$ are called precious thermocouples and others are base metal thermocouples. Precious thermocouples cannot be used without protection in reducing atmosphere. For temperatures in the range of 1500°C to 2200°C, special types of thermocouples made of W–(W–Re), Ir–W, Ir–(Ir–Rh) etc. are used. Size of the thermocouple wire is of some importance. Usually higher the temperature to be measured, the thicker should be the wire. As the size is increased, however, the time response of the thermocouple to temperature changes becomes low. Thermocouples may be prepared by twisting the two wires together and brazing, or preferably welding them as shown in Fig. 8.13. · The wires may be separated with the help of insulating beads or discs. The thermocouples can be used either in series or parallel. In a series combination the output is n times that obtained from a single thermocouple. Usually, they are combined in such a fashion that measurement junction appears like a point. **The series arrangement of thermocouples is called thermopile, and is used in the cluster form as detectors in balometers, or IR pyrometers. Parallel combination of thermocouples is used for averaging of temperature.**

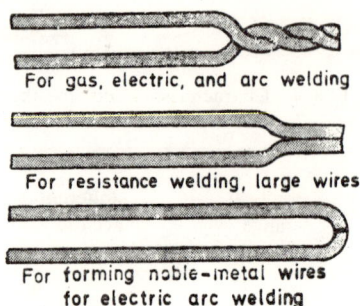

For gas, electric, and arc welding

For resistance welding, large wires

For forming noble-metal wires
for electric arc welding

Fig. 8.13   Common forms of thermocouple construction

Measurement of emf

The magnitude of the emf developed by a thermocouple is very small (in the range of 0.01 to 0.07 mV/°C) and thus its measurement requires sensitive devices. The measurement of the thermocouple output is done either by a millivoltmeter, or a voltage balancing potentiometer. Let a millivoltmeter of internal resistance R_i be used to measure the thermo emf E_s developed by a thermocouple. The thermocouple is connected to the millivoltmeter by lead wires having resistance R_l. The current i flowing in the circuit will be

$$i = \frac{E_s}{R_i + R_l + R_{th}},$$

where R_{th} is the resistance of thermocouple wires. The millivoltmeter will indicate a voltage V given by

$$V = R_i\, i = E_s - V\, \frac{R_l + R_{th}}{R_i}$$

The voltage indicated by the millivoltmeter will be lower from E_s by the voltage drop in the external circuit. In order to reduce the error in the temperature measurement lead wires of very low resistance and a millivoltmeter of high internal resistance should be employed.

A number of thermocouples can be connected to a single millivoltmeter through a selector switch at reference junction ; voltage developed by any thermocouple can be measured by connecting it to the millivoltmeter by the selector switch.

Often voltage balancing potentiometer, either manually or automatically operated, is used for the measurement of thermo emf. Since no current flows in the circuit at balance, resistance problems in the leads are largely eliminated. Figure 8.14 (a) shows a thermocouple connected to a potentiometer. Initially the switch is connected to position 2 so that the standard cell is in circuit. The resistance R_I is so adjusted that the galvanometer indicates a balance. At balance

$$E_{sc} = RI,$$

where E_{sc} is the emf of the standard cell, and I is the potentiometer current. Under this condition, current through the calibrated wire R_c is standardised. Now the thermocouple is brought in the circuit by switching to position 1. A balance position is obtained by adjusting the slide on the calibrated wire. If the resistance of the calibrated wire, for a particular position of the slide, is R_1, the thermo emf E_s is

$$E_s = IR_1 = E_{sc} \frac{R_1}{R}$$

Note that no current flows through the thermocouple circuit at balance.

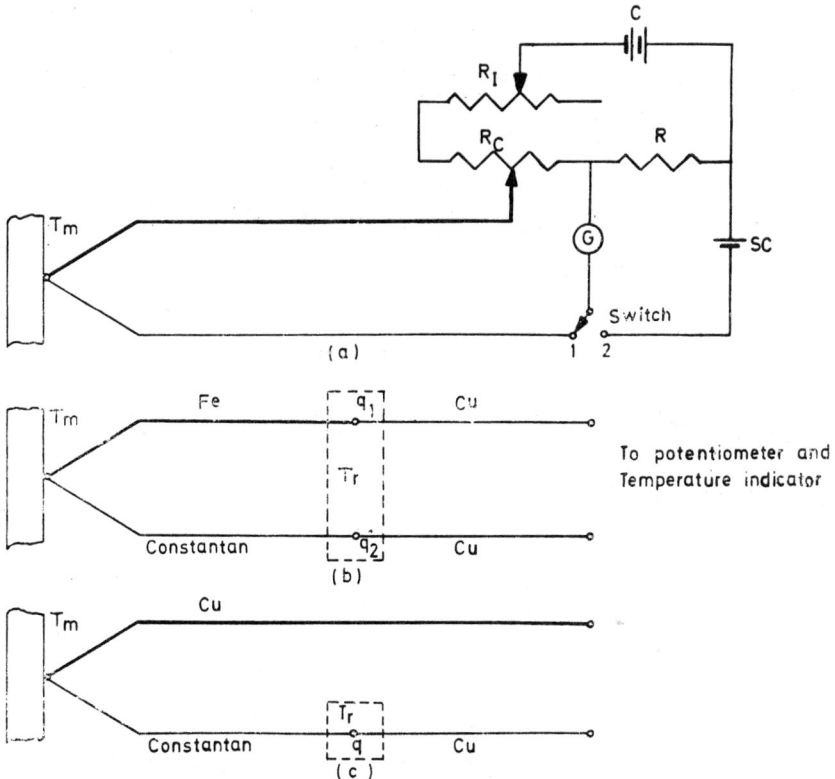

Fig. 8.14 Measurement of emf

SC = Standard cell

R = Fixed resistance

C = Cell

R_c = Calibrated resistance wire

R_I = Current adjustment resistance

G = Galvanometer

Further the accuracy of measurement depends on the constancy of current I, which should be checked time to time.

Thermocouple wires are relatively expensive compared to most common materials like copper. It is, therefore, desirable to minimise the use of more expensive material by using extension leads. The arrangements showing the use of lead wires are shown in Fig. 8.14 (b) and (c). As pointed out earlier, the temperature of reference junction should be maintained constant and be accurately known. Some baths are used in laboratory for maintaining the temperature of reference junction constant.

Thermocouple Burn-out Feature

Thermocouples are also used for controlling temperature in addition to measure it. As the temperature of measuring junction decreases its emf decreases. In a controller, the response to a decreasing emf signal is to increase heat input process. In the case of a burned-out thermocouple, no emf is generated creating the danger of maximum heat input to the process taking place.

In most controllers, the methods used to avoid excessive heat into the process are similar to the one explained in the Fig. 8.15. A power source P of voltage V is connected across the thermocouple. The resistance of thermocouple including the resistance of the connecting wires A and B is R_3. The voltage drop across the thermocouple and the lead wires due to current from P is

$$V_1 = \frac{VR_3}{R_2 + R_3}$$

This voltage will cancel the emf generated by the thermocouple. However, this effect can be taken into account in the calibration process itself.

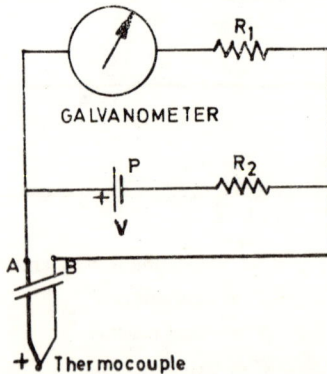

Fig. 8.15   Schematic of thermocouple burnout feature

Consider the situation when the thermocouple is burnt out. There is no thermo-emf and hence the bucking effect from P disappears. This results in a flow of current in the galvanometer circuit. The voltage drop caused by the current across the galvanometer and calibration spool (resistance R_1) is now $V_2 = \dfrac{VR_1}{R_1 + R_2}$. In practice, the design of circuit is so judicious that $V_2 > V_1$. On burn-out a voltage drop V_2 in place of V_1 is sensed and hence the controller will shut down the heat input. As a numerical example, consider $V = 1$ volt, $R_1 = 500\ \Omega$, $R_2 = 500\ \Omega$, $R_3 = 1\ \Omega$ then $V_1 = 0.2$ mV. This drop in voltage across the thermocouple $Fe-Ni_{40}\ Cu_{60}$ is equivalent to about 4°C change of temperature and thus compensated in calibration. Further $V_2 = 91$ mV, thus $V_2 > V_1$. The voltage V_2 is much larger than is obtained from the thermocouple and hence the controller will shut down the heat input.

Dynamics of thermocouple

A thermometer (thermocouple) can be considered as an element with mass m, specific heat c and area A. Let the heat transfer coefficient be h. The thermometer is inserted in a temperature field T_i. Then using a very simplified treatment, one has

$$hA\ (T_i - T) + mc\frac{dT}{dt} = 0.$$

or

$$\frac{dT}{T_i - T_i} = -\frac{hA}{mc}\ dt$$

on integration,

$$\ln (T_i - T) = -\frac{hA}{mc}\ t + C$$

where C is a constant. Applying the boundary condition that at $t = 0$, $T = T_0$, where T_0 is the temperature of the surroundings, one obtains,

$$\frac{T_i - T}{T_i - T_0} = e^{-(hA/mc)t}$$

or

$$T = T_i - (T_i - T_0)\ e^{-(hA/mc)t}$$

The response of the system is, therefore, first order with a time constant $\tau = mc/hA$.

In order to have fast response, the mass and specific heat must be small and the area and heat transfer coefficient should be large. It is, in general, said that the fast temperature changes cannot be followed fast enough. The frequency response depends on the time constant and is poor at high frequencies. Thus a compensation network is used to increase the frequency response of the thermocouple. The disadvantage of the compensation network is that it reduces the thermocouple output. A typical thermocouple compensation network is shown in Fig. 8.16. The thermocouple voltage is represented by E_l and the output voltage by E_0. The transfer function of the system is written as

$$\frac{E_0}{E_l} = \frac{Z_2}{Z_1 + E_2}$$

where Z_1 is the impendance due to R_c and C, and Z_2 due to R, respectively.

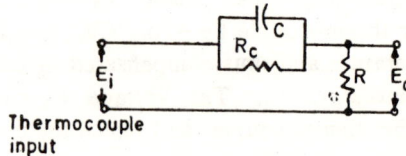

Fig. 8.16 Thermocouple compensation network

The impedances Z_1 and Z_2 are expressed as

$$Z_1 = \frac{R_c\,(1/CS)}{R_c + (1/CS)}, \text{ and}$$

$$Z_2 = R,$$

where $S =$ Laplace operator

  $= (i\omega)$ for frequency response.

If a sinusoidal input of frequency ω is given to the thermocouple compensation network, one obtains

$$\frac{E_0}{E_l} = a\frac{1 + i\omega\tau_c}{1 + i\omega a\tau_c}$$

where $a = R/(R + R_c) =$ steady state output, and

  $\tau_c = R_c C.$

The amplitude response is, therefore, given by

$$\left|\frac{E_0}{E_l}\right| = a\sqrt{\frac{1 + \omega^2\tau_c^2}{1 + \omega^2 a^2\tau_c^2}}$$

This particular network attenuates smaller frequencies more than high frequencies and has a response to step input which is approximately opposite to that of the thermocouple. High frequency compensation of the network is improved as the value of a is decreased. This brings about a decrease in the output.

Comparison between resistance thermometers and thermocouples

(i) Resistance thermometers are more accurate than thermocouples. They are usually rated at $\pm$ 0.3°C. Thermocouples always offer a slight possibility of inaccuracy due to changes in the reference junction temperature. The errors due to this could be $\pm 1.0°C$ or more.

(ii) Resistance thermometers have greater sensitivity than thermo-couples.

(iii) Earlier the response of resistance thermometer element was slower than that of thermocouple. Response of modern resistance elements is about the same as that of a thermocouple.

(iv) Thermocouples require a somewhat more frequent replacement. But thermocouples are cheaper as compared to resistance thermometer elements. Thus final cost of either arrangement is about the same. The inconvenience caused in replacing the thermocouple might throw the balance in favour of resistance thermometers.

(v) In general, the resistance thermometer is used in preference to the thermocouple wherever possible. The principal limitation is temperature.

8.6 Pyrometry

So far temperature measuring devices which are placed in contact with the body whose temperature is to be measured have been considered. If the body is very hot, the contact of the thermometer with the body is likely to damage it completely. In other cases a hot body may not be accessible for contact measurement. In such situations, one relies on instruments which do not require to be placed in contact and the distance between the source and the instrument does not affect the measurement. These instruments are called pyrometers and are based on radiation thermometry. There are two distinct types of pyrometers: total radiation pyrometers, and optical pyrometers.

Total radiation pyrometer, as the name implies, accepts a controlled sample of total radiation, and through determination of the heating effect of the sample, obtains a measure of temperature. Optical pyro-meters are again of two types: Photon flux, and monochromatic. In one the output from a photosensor indicates the temperature while in the other spectral intensity is compared with a calibrated standard. Before discussing theory and working of pyrometers, a little of the fundamentals of radiation is discussed here.

Radiation Fundamentals

All bodies above absolute zero radiate energy; this radiation depends upon its temperature. The ideal thermal radiator is called a black body. Such a body would absorb completely any radiation falling on it. The relation between black body radiation, temperature and wave-length is described by Planck's equation:

$$W_\lambda = \frac{C_1 \cdot}{\lambda^5(e^{C_2/\lambda T} - 1)},$$

where W_λ = hemispherical spectral radiant intensity W/cm² μm

$C_1 = 37,413$ W μm⁴/cm² (constant),

$C_2 = 14,388$ μm K (constant).

λ = wavelength in μm and

T = temperature in K.

The quantity W_λ is the amount of radiation emitted from a flat surface into a hemisphere, per unit wavelength, at the wavelength λ. Figure 8.17 illustrates the variation of W_λ with λ for various values of temperature. The W_λ vs λ curves exhibit peaks at a particular wavelength. The peak shifts towards higher wavelengths as the temperature decreases. The maximum of W_λ vs λ curve is obtained by letting

$$\frac{dW_\lambda}{d\lambda}\bigg|_{\lambda=\lambda_p} = 0.$$

This gives $\lambda_p T$ = constant = 2,891 μm K. This relations is known as Wien's displacement law.

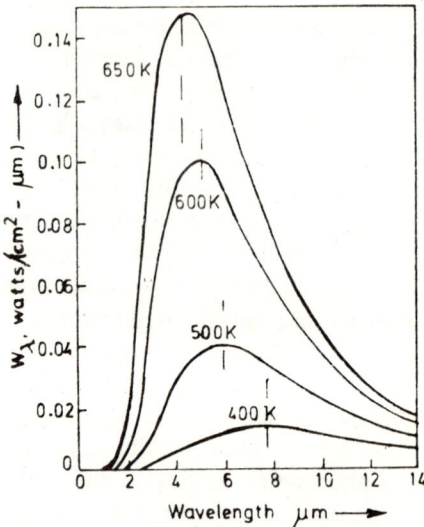

Fig. 8.17   Black body radiation

Further the area under the curve gives the total energy emitted by a black body at a particular temperature and is given by

$$W_t = \int_0^\infty W_\lambda \, d\lambda \quad \sigma T^4 \ (\text{W/cm}^2).$$

This is the total power radiated by a flat surface into a hemisphere. The above relation is called Stefan-Boltzmann law; σ is the Stefan's constant and its value is

$$\sigma = 5.67 \times 10^{-12} \ \text{W/cm}^2/\text{K}^4$$

While the concept of a black body is a mathematical idealisation it is possible to construct a real physical body whose behaviour approximates closely to that of a black body. The deviations from black body radiation are expressed in terms of emittance of the real physical body. Let the actual hemispherical spectral radiant intensity of a real body at temperature T be $W_{\lambda a}$. It is assumed that $W_{\lambda a}$ can be measured. Then the hemispherical spectral emittance $\epsilon_{\lambda,T}$ is defined as $\epsilon_{\lambda,T} = W_{\lambda a}/W_\lambda$. Thus for $W_{\lambda a}$ one can write

$$W_{\lambda a} = \frac{C_1 \epsilon_{\lambda,T}}{\lambda^5 \left(e^{C_2/\lambda T} - 1\right)}$$

Similarly, the total power W_{ta} of a an actual body is given by

$$W_{ta} = \int_0^\infty \epsilon_{\lambda,T}\, W_\lambda\, d\lambda$$

and assuming that W_{ta} can be measured, the total hemispherical emittance $\epsilon_{t,T}$ is

$$\epsilon_{t,T} = \frac{W_{ta}}{W_t}$$

Thus if $\epsilon_{t,T}$ is known, the total power radiated by a real body at temperature T is

$$W_{ta} = \epsilon_{t,T}\, \sigma T^4$$

One can ascribe a black body temperature T' to any real body by $\sqrt[4]{\epsilon_{t.T}}\, T = T'$ such that

$$W_{ta} = \sigma T'^4.$$

Therefore, any real physical body is ascribed by $\epsilon_{\lambda,T}$ and $\epsilon_{t,T}$ so that its radiation characteristics are same as that of a black body at a particular temperature.

A body whose $\epsilon_{\lambda,T}$ is independent of wavelength at a temperature T is called a grey body. In this case, $\epsilon_{\lambda,T} = \epsilon_{t,T}$. Since many radiation thermometers operate in a restricted band of wavelengths, the hemispherical band emittance $\epsilon_{b,T}$ has been defined as

$$\epsilon_{b,T} = \frac{\int_{\lambda_a}^{\lambda_b} \epsilon_{\lambda,T}\, W_\lambda d\lambda}{\int_{\lambda_a}^{\lambda_b} W_\lambda d\lambda}$$

This is just the ratio of the total powers, of actual and black bodies, within the wavelength band λ_a and λ_b when they are at temperature T. If the actual power in the wavelength band λ_a and λ_b can be measured directly, $\epsilon_{b,T}$ can be found without knowing $\epsilon_{\lambda,T}$. For grey bodies $\epsilon_{b,T} = \epsilon_{\lambda,T}$.

UNCHOPPED (D.C.) BROAD BAND RADIATION THERMOMETERS

These instruments use blackened thermopile or bolometer as detector and focus the radiation by means of either lenses or mirrors. The lenses have, however, selective transmission and have wavelength dependant focal lengths, unless specifically designed. The use of mirrors rather than of lenses is an attempt to eliminate some of these problems. Figure 8.18 shows two arrangements, one is a lenses based configuration and other is a mirror based configuration. The imaging system makes an image of an area of the target which is at a temperature T_1 on the detector. The spatial resolution of the temperature distribution at the target is governed by the optical design considerations and detector size. The focussed radiation raises the temperature of the measuring junction (assuming thermocouple is used as a detector) until conduction, convection and radiation losses just balance the heat input. Usually the temperature difference between the temperature of the junction and of its surroundings is less than 40°C even if the target is incandescent. Following the heat balance equation i.e. heat loss = radiant heat input, an over-simplified analysis gives

$$A (T_2 - T_3) = B T_1^4,$$

where A, B are constants, T_2 and T_3 are the temperatures of junction and its surroundings respectively [$(T_2 - T_3) < 40°C$]. For smaller temperature difference $(T_2 - T_3)$, the emf developed by the thermocouple may be taken proportional to $(T_2 - T_3)$. The voltage output is therefore proportional to T_1^4. For targets at high temperatures, the results of temperature measurement are quite good. The temperature T_2 and T_3 and consequently $(T_2 - T_3)$ are influenced by the environmental temperature.

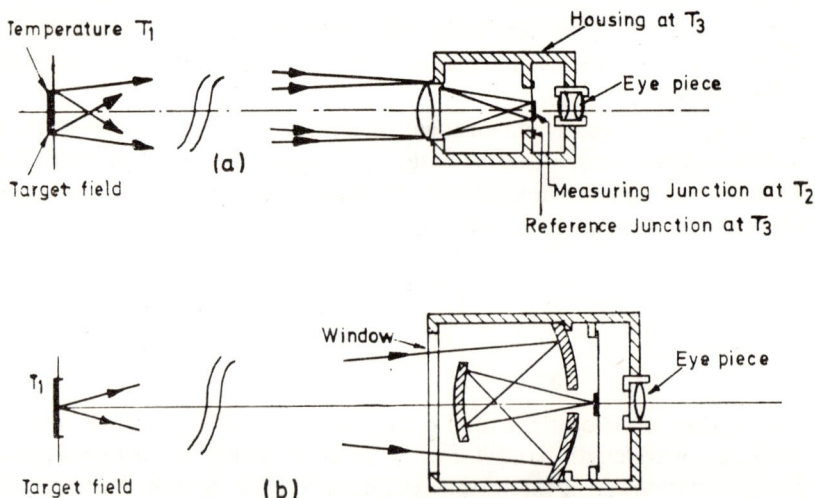

Fig. 8.18   (a)   Refracting arrangement
          (b)   Reflecting arrangement

Thus the housing temperature is thermostatically controlled to eliminate the influence of the variation of ambient temperature on $(T_2 - T_3)$. Calibration of the instrument is generally independent of the target distance, as long as the target fills the field of view. For targets at near distances, focussing is required which may upset the calibration.

Instead of a thermocouple, a thermopile may be used as a detector to obtain proportionately higher output. The design of a thermopile is shown in Fig. 8.19. All junctions are so arranged as to occupy a very small area, since a focussed beam falls on it. The other ends (reference junctions) are fixed to a massive support. The thermopile may have 1 or 2 to 20 or 30 junctions. A thermopile with a few junctions has a faster response due to less mass, but low sensitivity. Dynamic response of these systems is roughly first order with time constants ranging from 0.1 sec. to 2 sec. The instruments are available to measure temperatures down to $-18°C$ giving negative voltage output. Theoretically, there is no limit to the temperature which can be measured with these types of instruments. Commercial instruments are available which measure temperature as low as $-18°C$ and as high as $1760°C$.

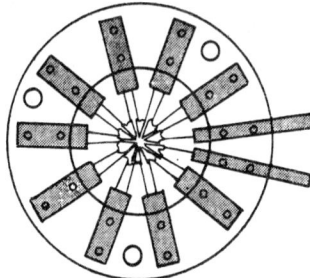

Fig. 8.19 A thermopile

CHOPPED (A.C.) BROAD BAND RADIATION THERMOMETERS

The d.c. radiation thermometers are relatively less sensitive and require ambient temperature compensation. If high sensitivity is required for the same detector type, the output must be amplified considerably. It has been pointed out earlier that it is easier to construct a high gain a.c. amplifier than its d.c. counterpart. Therefore many infrared systems employ chopping of the radiation falling at the detector. Further, any unwanted d.c. signal can be eliminated by tuning the amplifier to the chopper frequency. Ambient temperature and reference junction compensation may also be obtained. Figure 8.20 shows a schematic of an a.c. broad band radiation thermometer. The target is imaged at the detector with the help of a mirror. A blackened chopper is mounted in front of the detector. The chopper periodically cuts off the radiation falling on the detector from the target. The detector 'sees' the radiation alternately from the chopper and the target. For the measurement of temperature of a target at high temperature, sufficient accuracy is achieved by thermostatically controlling the chopper temperature. The effective use of chopping requires detectors of a very fast response. Therefore fast response bolometers or thermistors instead of thermocouples/thermopiles are used as radiation sensors. A typical instrument uses a square thermistor detector and has a field of view of $\pm 0.50°$. The chopping frequency is 180 Hz and an overall time constant of 8 ms. The temperature range is from ambient temperature to 1300°C and the distance range is from 50 cm to infinity without loss of calibration.

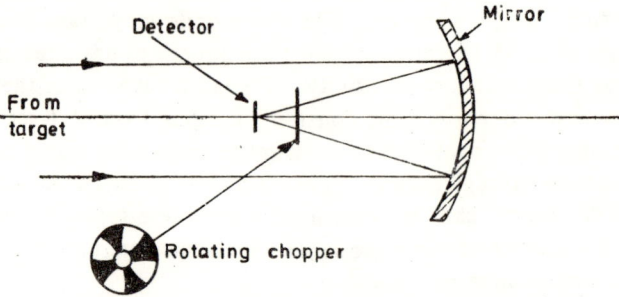

Fig. 8.20   Chopped radiation thermometer

The calibration of total radiation pyrometers is done with black body radiation. The output is proportional to T^4. When the pyrometer is employed to measure the temperature of a real body, its emissivity must be known. If the emissivity is not accurately known, the measured temperature value will be in error. The magnitude of the error can be obtained as follows: The output of the thermometer is proportional to T^4 and may be expressed as

$$e = K \epsilon T^4$$

where K is a constant.

On differentiation, one has

$$\frac{dT}{T} = \frac{1}{4} \frac{d\epsilon}{\epsilon}$$

Thus a 10% error in the value of emissivity will result only in 2.5% error in the measurement of temperature.

8.7   Optical pyrometers

Chopped (a.c.) Photon Flux Radiation Thermometers

The photon detectors (photocells, photo-transistors, or photo-multiplier tubes etc.) have a faster response. Their use reduces the effect of ambient temperature as they measure the photon flux rather than temperature. The photon flux measurement is independent of temperature variations so long the responsivity of the detector does not vary with temperature.

The relation between the hemispherical spectral photon flux N_λ (photon/(cm²–sec–μm) and temperature is expressed as

$$N_\lambda = \frac{2\pi c}{\lambda^4(e^{C_2/\lambda T} - 1)},$$

where c is the velocity of light. The N_λ vs λ curve for any temperature exhibits a maximum which is obtained by setting

$$\frac{dN_\lambda}{d\lambda} = 0.$$

It is given by

$$\lambda_{p,p} T = 3,669 \ \mu\text{m K}.$$

where $\lambda_{p,p}$ is the wavelength corresponding to the $N_\lambda \big|_{\text{max}}$ at temperature T. The maximum of the photon-flux occurs at a different wavelength than that of the radiant intensity. The total photon flux N_t can be obtained by integrating N_λ over all the wavelengths, i.e.

$$N_t = \int_0^\infty N_\lambda \, d\lambda = 152 \times 10^{11} T^3 \ \text{photon/(cm}^2\text{-sec)}$$

$$= \sigma_p T^3$$

The total photon flux varies as cube of the absolute temperature of the radiating body.

A schematic arrangement of a pyrometer using a photon detector and a chopper for the temperature measurement is given in Fig. 8.21.

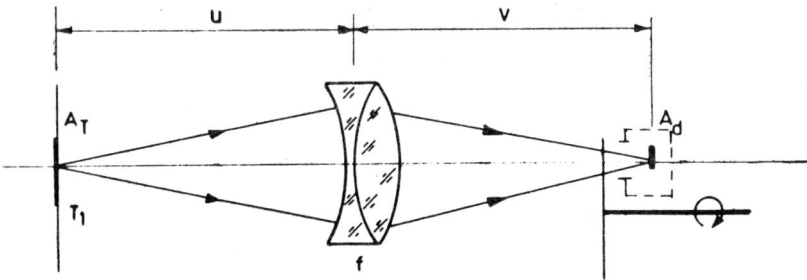

Fig. 8.21 Chopped photon flux thermometer

The thermometer uses a lens optics and makes an image of a target of area A_t on a detector of area A_d. It is assumed that the target image fills the detector area. Therefore, from the imaging and magnification conditions, one has

$$\frac{1}{u} + \frac{1}{v} = \frac{1}{f},$$

and

$$A_t = A_d \left(\frac{u}{v}\right)^2,$$

where f is the focal length of the lens, and u and v are the target and detector distances respectively as measured from the lens. Usually $u \gg f$, i.e. the thermometer looks at distant targets, and therefore, its image is formed near its focus, thus

$$A_t = A_d \left(\frac{u}{f}\right)^2.$$

The focal length of the lens, therefore, determines the spatial resolution of the instrument, i.e. the minimum area of the target whose average temperature is being measured. The photon-flux incident on the lens is given by

$$N_L = N_t A_t \frac{(\pi D^2/4)}{\pi u^2},$$

where

$$N_t = \int \epsilon_{\lambda, T} \, N_\lambda d\lambda$$

$$= \epsilon \int N_\lambda d\lambda \text{ for grey bodies}$$

$$= \epsilon \, E \, (T_t)$$

$E(T_t)$ is a function of target temperature alone. D is the diameter of the lens, therefore, $(\pi D^2/4)/u^2$ is the solid angle subtended by the lens aperture at the target. Assuming lens transmittance as τ_L, the number of photons on the other side of the lens is $\tau_L A_t \dfrac{D^2}{4u^2} \epsilon \, E \, (T_t)$. The photon-flux is brought to focus on the detector of area A_d by the lens. Therefore the photon flux density on the detector is

$$N_d = \tau_L \frac{A_t}{A_d} \frac{D^2}{4u^2} \epsilon E \, (T_t)$$

$$= \tau_L \frac{D^2}{4f^2} \epsilon E \, (T_t)$$

$$= \frac{\tau_L}{4} F^2 \epsilon E \, (T_t),$$

where $F \, (= D/f)$ is the F-number of the lens. Assuming that the detector output is proportional to N_d, it is seen that the sensitivity depends on the F-number of the lens. Since resolution depends on f and sensitivity on F, a judicious choice of D and f is made in the design of the optical system. For most of the detectors the relation between N_d and its output is linear. The overall output of the instrument can be expressed

$$e = K_d N_d \, m,$$

where K_d is the detector responsivity (V/photon/cm^2-s) and m is the amplifier gain. Substituting for N_d, one has

$$e = \left[\frac{\tau_L K_d m}{4} F^2\right] [\epsilon E \, (T_1)].$$

The term in the first bracket is a' constant of the instrument. The second term depends on the emissivity and temperature of the target. As long as the target image fills the detector area, the output is independent of the target distance. The variation of $E(T_t)$ with T for a black body target can be found by experimental calibration of the overall instrument. This is necessitated because the transmission function of the lens is wavelengths dependent, and the instrument uses a definite broad-band for its operation. It has been shown theoretically that the total photon-flux varies as T^3, and hence the output voltage e being proportional to flux also varies as T^3. Thus for grey bodies

$$e = K\epsilon T^3,$$

where K is a constant. Since the instrument is calibrated against black body radiation, the value of ϵ must be known to find the temperature of the non-black bodies. If the value of ϵ is in error, temperature will also be in error. The temperature error dT is given by

$$\frac{dT}{T} = \frac{1}{3}\frac{d\epsilon}{\epsilon}$$

Thus a 10% error in ϵ would result in a 3.3% error in the temperature.

An instrument of this class which uses mirror optics has a temperature range of 38°C to 1150°C, focussing range from 1 m to infinity, total field of 0.5° and output signal of 10 mV full scale. For the study of rapid temperature transients, the chopper may be turned off, thus the unchopped beam falls on the detector. The system has an overall time constant in the range of ms. Another instrument of this class accepts radiation in the wavelength band of 4.8 to 5.6 μm and obviously uses a mirror optics. This band is chosen particularly to avoid the effect of absorption bands of H_2O and CO_2 on the instrument's response. The instrument has a temperature range of 38°C to 540°C, time constant of 0.2 to 0.5 sec., focussing range from 0.5 m to infinity and temperature accuracy of 0.25% of the range or 1.5°C which ever is larger.

MONOCHROMATIC OPTICAL PYROMETER

This pyrometer is the most accurate of all the radiation pyrometers and is used as calibration standard above the gold point.

A schematic of a disappearing filament type optical pyrometer is given in Fig. 8.22. A lens makes an image of the target area on a flat tungsten filament lamp. A calibrated absorption filter is inserted in front of the filament to extend its range. The image of the target and filament are viewed by an eye-piece or a low power microscope. A red filter which passes a band of wavelengths around 0.65 μm is placed between the filament and the eye-piece for assisting in the visual matching of the field. The current flowing in the filament circuit determines its brightness temperature. The tungsten lamp is first calibrated by visually comparing the brightness of the image of a black body source of known temperature with that of the tungsten filament. The observer controls the current until filament disappears in the superposed field. At this condition brightness of the target image and that of the lamp are equal. A relation between filament current i and black body temperature T is a valid calibration curve under the conditions prevailing during calibration process. The fields of view as occurring at mismatch and null conditions are also shown in Fig. 8.22.

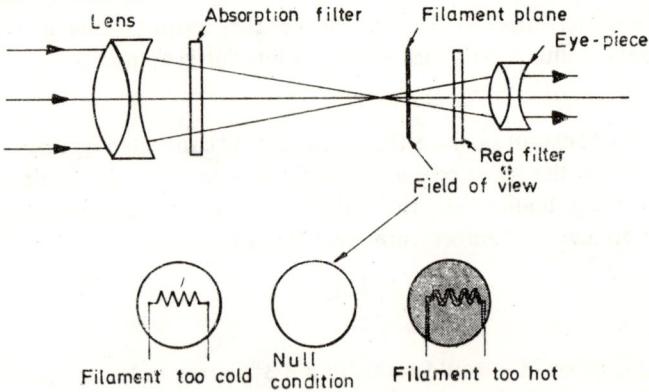

Fig. 8.22   Schematic of an optical pyrometer

There is theoretically no upper limit of temperature that can be measured by the optical pyrometer. Practically, however, the lamp cannot be operated above a certain current or brightness value because of its long term stability and life.   However, the range is extended by inserting absorption filters before the filament to attenuate the brightness of the image of target. These filters are initially calibrated. A greater emphasis is placed on the stability of the lamp and hence the rated filament current should not be exceeded. A tungsten lamp can be used upto 200 hours before recalibration is required.

The working of an optical pyrometer is based on the principle that at a given wavelength λ, the radiant intensity (brightness) of a black body varies with temperature T as

$$W_\lambda = \frac{C_1}{\lambda^5 \{\exp (C_2/\lambda T) - 1\}}.$$

The lamp is calibrated against a black body, so that the brightness of the lamp at wavelength λ is also given by this equation. Therefore, when the brightness of the target image matches with that of the filament, one has

$$\frac{C_1}{\lambda_0^5 \{\exp (C_2/\lambda_0 T_L) - 1\}} = \frac{\epsilon_{\lambda_0, T} C_1}{\lambda_0^5 \{\exp (C_2/\lambda_0 T) - 1\}}$$

where T_L is the lamp temperature, $\epsilon_{\lambda_0, T}$ is the emissivity of the body at mean wavelength λ_0 and T is the temperature of the body. λ_0 is the mean wavelength of the filter, i.e. $\lambda_0 = 0.65$ μm. For temperatures less than 4000°C, the term exp $(C_2/\lambda_0 T)$ is much greater than unity. Hence one has

$$\exp\left(-C_2/\lambda_0 T_L\right) = \epsilon_{\lambda_0,\,T}\exp\left(-C_2/\lambda_0 T\right)$$

or

$$\frac{1}{T} - \frac{1}{T_L} = \frac{\lambda_0}{C_2}\log_e\left(\epsilon_{\lambda_0,T}\right)$$

If the target is a black body $(\epsilon_{\lambda_0,\,T} = 1)$, $T = T_L$ and hence there is no error in the temperature measurement. However, if $\epsilon_{\lambda_0,\,T} < 1$ but known, this equation can be used to obtain T in terms of T_L. If the value of $\epsilon_{\lambda_0,\,T}$ is not known, the percentage error in temperature is given by

$$\frac{|dT|}{T} = \frac{T\lambda_0}{C_2}\left(\frac{d\epsilon}{\epsilon}\right)$$

If the target is at 1200 K, a 10% error in the value of ϵ results only in a 0.54% error in T. Therefore, the errror caused by the inexact value of ϵ for a particular target is not as great for an optical pyrometer as for an instrument using a wide band of wavelengths like photon-flux or radiation pyrometer. Further, the target emittance need be known only at one wavelength λ_0. The limitation of the optical pyrometer is that it can be used to measure temperatures above 700°C since it requires a visual brightness match by a human operator. If ϵ is known exactly the temperature can be measured with optical pyrometers with an error of

3°C at 1000°C,

6°C at 2000°C,

and
40°C at 4000°C.

TWO-COLOUR RADIATION THERMOMETER

Even in an optical pyrometer, the emittance of the target should be known at one wavelength. If the body is distantly away and inaccessible, it may be difficult to measure its emittance. The inaccurate values of emittance result in error in the temperature measurement by all types of radiation thermometers. There has been no permanent solution to this problem, but the use of two wavelengths to measure temperature has found some support. If it is assumed that the target is a grey body, its emittance need not be known. In an otherwise situation, two wavelengths are taken fairly close so that the emittance may be assumed constant.

The two wavelengths method requires that W_λ be measured at two different wavelengths and their ratio be used to measure temperature. If the target temperature is less than 4000°C, the term exp $(C_2/\lambda T)$ is much greater than 1.0, and hence the radiant intensities W_1 and W_2 at wavelengths λ_1 and λ_2 respectively can be expressed with a close approximation as

$$W_1 = \epsilon_{\lambda_1} C_1 \lambda_1^{-5} \exp\left(-C_2/\lambda_1 T\right)$$

and

$$W_2 = \epsilon_{\lambda_2} C_1 \lambda_2^{-5} \exp\left(-C_2/\lambda_2 T\right)$$

Therefore, their ratio is expressed as

$$\frac{W_1}{W_2} = \left(\frac{\epsilon_{\lambda_1}}{\epsilon_{\lambda_2}}\right)\left(\frac{\lambda_1}{\lambda_2}\right)^{-5} \exp\left\{-\frac{C_2}{T}\left(\frac{1}{\lambda_1} - \frac{1}{\lambda_2}\right)\right\}$$

For grey bodies $\epsilon_{\lambda_1} = \epsilon_{\lambda_2}$, and hence

$$\frac{W_1}{W_2} = \left(\frac{\lambda_1}{\lambda_2}\right)^{-5} \exp\left\{-\frac{C_2}{T}\left(\frac{1}{\lambda_1} - \frac{1}{\lambda_2}\right)\right\}$$

Thus one finds that the ratio W_1/W_2 is independent of emittance for (i) for grey body, and (ii) as long as the value of emittance is numerically the same at λ_1 and λ_2. In one commercial instrument, two filters are mounted on a rotating wheel. The incident beam is alternately passed through each of the filters on its way to the photo-detector. The ratio W_1/W_2 is obtained by electronic processing of the signal from the detector. Commercial instruments which cover a temperature range from 1400°C to 4000°C are available.

8.8   Quartz crystal thermometer

The resonant frequency of a quartz crystal plate depends on the temperature. When the crystal is cut at a proper angle, the change in the frequency is linearly related to the change in the temperature of the crystal. This linear correspondence has been utilised in a very noval and highly accurate way for temperature measurement. Since the frequency of a resonant cavity is measured, the temperature measurement is not succeptible to any noise pick up in the connecting cable etc. The commercial models of the device utilise electronic counters and digital readouts for the frequency measurement. Sensitivities in the range of 10^{-3}°C are achievable with these devices.

8.9   Microwave techniques of high temperature measurement

In this instrument again the frequency is measured and the output is digital. The resonant frequency f_r of a cavity is given by

$$f_r = \frac{c}{2}\left[\left(\frac{\lambda}{L}\right)^2 + \left(\frac{U_{nm}}{\pi r}\right)^2\right]^{\frac{1}{2}},$$

where U is the argument for which Bessel function is zero, r is the radius and L the length of the cavity. The cavity expands or contracts with temperature and hence the resonant frequency changes.

Exercises

1. A bimetallic strip of length 125 mm is made by joining two strips of brass and Monel at 45°C. The thickness of the strip of brass ($\alpha = 2.02 \times 10^{-5}/°C$, $E = 0.95 \times 10^{11} \text{ N/m}^2$) is 0.35 ± 0.005 mm and that of Monel ($\alpha = 1.35 \times 10^{-5}/°C$, $E = 1.9 \times 10 \text{ N}^{11}/\text{m}^2$) is 0.25 ± 0.003 mm. Calculate the deflection sensitivity (deflection per degree Celsius of temperature difference). Estimate the uncertainty in the deflection sensitivity.

2. A bimetallic strip of dimensions $150 \times 12 \times 1.0 \text{ mm}^3$ is made by joining two strips of same thickness, one of stainless steel ($\alpha = 1.73 \times 10^{-5}/°C$, $E = 1.93 \times 10^{11} \text{ N/m}^2$) and other of iron ($\alpha = 1.08 \times 10^{-5}/°C$, $E = 0.91 \times 10^{11} \text{ N/m}^2$). Calculate (i) the motion of the free end per unit temperature change (mm/°C), and (ii) the force required to hold the strip such that deflection is zero when the temperature increases by a unit amount.

3. The specific volume V of mercury varies with temperature T according to the relation $V = V_0 (1 + aT + bT^2)$ where $a = 0.1818 \times 10^{-3}/°C$ and $b = 0.0078 \times 10^{-6}/°C^2$. A high temperature thermometer is constructed of a Monel tube having an inside diameter of 0.8 ± 0.004 mm. After evacuating the tube, mercury is filled in the tube such that a column of 100 ± 0.2 mm is achieved when the thermometer temperature is 260°C. What is the uncertainty for a temperature of 260°C measured by this tube if the uncertainty in the height measurement is ± 0.20 mm?.

4. A platinum thermo-resistive element is to be used in the bridge circuit of Fig. 8.23 to measure temperatures in the range of 200°C to 300°C. The potentiometer P is used to adjust the output to zero at 0°C. All the resistance have a temperature coefficient of $1.0 \times 10^{-5} \Omega/°C$. It is desired to measure the temperature to a precision of 0.1°C. Determine (i) the supply voltage and (ii) the value of P to ensure zero adjustment. What range and resolution must the meter have?. How large a temperature variation can be tolerated in the bridge circuit? Following are the characteristics of the platinum thermo-resistive element:

$$\text{range} = -200°C \text{ to } 500°C$$
$$\text{sensitivity} = 0.1 \ \Omega/°C \text{ at } 0°C$$
$$\text{resistance} = 25 \ \Omega \text{ at } 0°C$$
$$\text{maximum current} = 10 \text{ mA}.$$

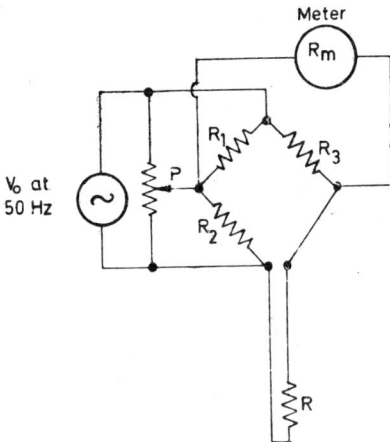

$R_1 = R_2 = 250 \pm 2\%$
$R_3 = 25 \pm 2\%$
$R_m = 100 \pm 2\%$

Fig. 8.23

5. A thermistor element with $\beta = 3420\ K$ is used to measure the temperature. The resistance of the thermistor at $100°C$ is known to be $1050 \pm 5\Omega$. Find the temperature and its uncertainty when its resistance is measured to be $2400 \pm 7\Omega$.

6. An iron-constantan couple is used to measure a temperature of $500°C$. The two reference junction are attached to a copper block to maintain their temperature equal at the room temperature. The meter is connected via copper leads. How much error will be caused in the measurement if the room temperature varies from $20°C$ to $27°C$?

7. What surface temperature of a black body is needed to radiate $120\ W/m^2$.

8. Estimate the total power found above the wavelength of $10\ \mu m$ for black body radiation at $T = 500°K$.

9. When a target of spectral emissivity ϵ_λ is viewed with a monochromatic pyro-meter, the measured temperature is somewhat less than the true temperature and will depend on the wavelength at which the measurements are made. The error is given by

$$\frac{T - T_e}{T} = \frac{\lambda T_e}{C_2}\ \log \epsilon_\lambda$$

where $T =$ true temperature, and $T_e =$ measured temperature. For the measure-ment at $0.615\ \mu m$ and $T = 1350°C$, calculate $(T - T_e)$ as a function ϵ_λ.

Optical Methods of Measurements

9.1 Introduction

Although a number of optical methods for measurements have been discussed in some chapters, only those specialised methods which have been developed after the emergence of the laser have been presented in this chapter. Laser is bestowed with a number of characteristics which make it unique for some measurements. The laser light is characterised by the following four properties:

   (i)   High intensity
   (ii)   Directionality
   (iii)   Coherence, and
   (iv)   Monochromaticity

The properties which are relevent to the measurement, particularly, of process variables are the last three: high intensity is usually associated with thermal effects which are of no relevance to the discussion of this chapter.

Laser can oscillate into a number of longitudinal frequencies, i.e. its radiation may comprise of a number of discreet frequencies. It may also have a number of transverse modes. In most of the measurements it would be assumed that the laser oscillates in a fundamental transverse electromagnetic mode denoted as TEM_{oo} mode and have a single frequency output. The irradiance distribution in the laser beam oscillating in TEM_{oo} mode is of gaussian spatial profile, and this distribution is maintained as the beam progresses in space. In fact TEM_{oo} mode is the fundamental mode which includes the maximum energy in the diffraction cone as it propagates.

9.2 Laser beam as a light pointer

Figure 9.1 illustrates a schematic of a very simple device that can be used for linear displacement measurement, or for the control of sheet thickness. The light is reflected off the face of a prism which can rock up and down when the plunger moves up and down. The reflected light suffers an angular shift that, due to the long distances involved, is greatly amplified on the screen. It is indeed possible to devise the instrument using a beam from an incandescent lamp. The laser offers an extreme ease of operation. Due to the diffraction limited collimation, the laser beam is often called an 'optical wire'. Further the center of the beam can be located with ease. Although the beam may expand due to its inherent divergence on propagation, beam profile is maintained and hence the center of the beam can be located with reasonable accuracy.

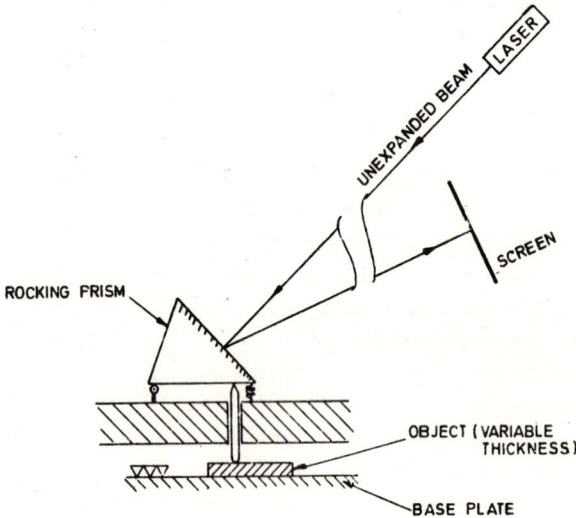

Fig. 9.1   Schematic for displacement measurement

If still higher accuracy is desired, a quadrant detector is used. In the quadrant detector the photo-sensitive area is divided into four quadrants. Each of the quadrants produces a signal. These singals are added in such a way that the indicator shows zero when the beam falls symmetrically about the point of intersection of the quadrants. For a laser beam of a small diameter, a very small relative displacement between the detector and the beam causes a relatively large imbalance output signal. For large beam diameters the sensitivity is not so high. An enhanced sensitivity can be obtained by using a diffracting aperture in the beam before the detector. Only when the diffracting aperture is symmetrically located with respect to the beam, the diffraction pattern is symmetrical. Very small displacement of the aperture results in a large asymmetry in the diffraction pattern.

In practice a beam from a He–Ne laser of about 2mW capacity without any beam expansion can be used over distances of a few meter. Using quadrant detectors with a diffracting aperture, an accuracy of $\pm$ 10 μm in the beam displacement is possible, and hence plunger displacement smaller than 0.1 μm can be sensed.

9.3 Length/displacement measurement

The length/displacement is measured in terms of wavelength of light. Earlier wavelength of the orange-red light of Kr[86] lamp was accepted as a standard for length. Because of operational difficulties with its use 1.15 μm and 0.6328 μm transitions of a He–Ne laser are being introduced as standard of length. The wavelength stability of a He–Ne laser ranges from 10^{-6} to 10^{-12}. For reasons of control on environmental conditions, a wavelength stability of 1 part in 10^7 is adequate for moderate displacement measurement.

Length measurement is done with the help of an interferometer that provides an amplification in excess of 4000. The basic principle of interferometric measurement can be understood with reference to Fig. 5.17. When the interferometer is properly aligned, a displacement of half-wavelength of a mirror results in the shift of a fringe pattern by one fringe width. A photo-detector is used to sense the irradiance variations occuring in the fringe field, and the displacement can be obtained by counting the number of maxima occuring in the detector output when the mirror is displaced. There are many other schemes available for the length measurement which are superior to the brute force method outlined above.

However for an interferometric configuration to be acceptable for displacement measurement with a laser beam, a number of requirements should be met. Some of them are:

(i) The laser should have adequate long term stability to suit the displacement measurements under existing environmental conditions with the expected accuracy. This requirement can be met easily for short displacements with equal-path interferometer.

(ii) The lay-out of the optical elements in the configuration should be such that no part of the laser beam is retro-reflected back in the laser. Interaction of the reflected beam with that generated inside the cavity may cause violent amplitude and frequency fluctuations.

(iii) The moving mirror should not tilt or displace the beam when it is being translated thereby requiring highly accurate guideways which are prohibitively expensive. For this reason, the mirrors in the interferometer are replaced by corner-cubes.

(iv) It should be possible to apply corrections for the change in environmental variables when the interferometer is in use.

(v) It should also be possible to know the direction of displacement. The direction discrimination is provided by sensing signals in quadrature and processing them further.

A schematic of an interferometer that meets many of these requirements is shown in Fig. 9.2(a). A collimated beam polarised linearly at 45° is split into two beams which propagate to corner cubes C_1 and C_2. The reflected beam from C_2 traverses through a $\lambda/4$ plate and hence becomes circularly polarised. The beam from C_1, on retro-reflection, meets the beam from C_2 at the beam splitter. These two beams, one circularly polarised and the other linearly polarised at 45°, on passage through a Wollaston prism are decomposed further into two beams each and are angularly separated. One set of beams produces $\sin^2$ and other $\cos^2$ type of irradiance distribution. These two fringe patterns are sensed by two independent photo-detectors. Therefore two signals in quadrature are available to provide direction discrimination.

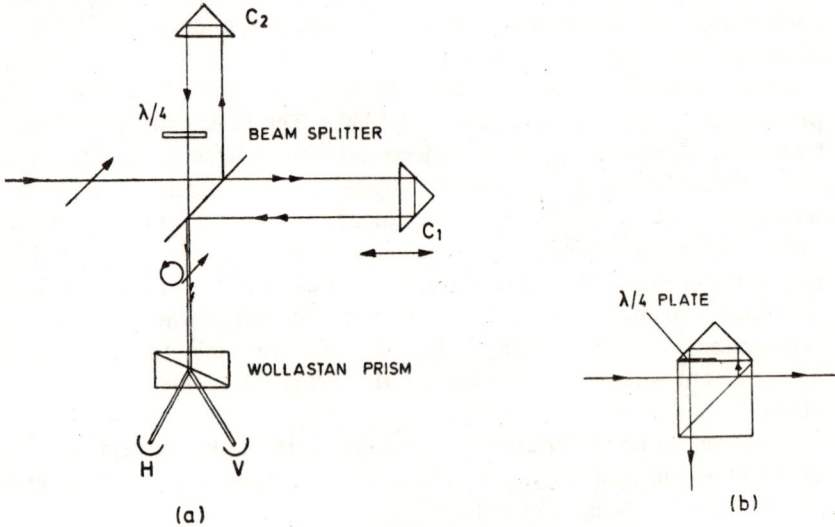

Fig. 9.2   (a)   Schematic of a laser interferometer
            (b)   Beam splitting assembly

In practice the beam splitter in the form of a cube, $\lambda/4$ plate and corner cube C_2 are assembled as a single unit as shown in Fig. 9.2(b). A block diagram of the signal processing circuitry and display is given in Fig. 9.3. The sinusoidal signals from the photo-detectors are shaped to rectangular wave forms which are represented by two states 1 and 0. The sequence is examined in a logic circuit and a counting command signal is obtained and fed to the counter. It is possible to produce one, two or four pulses per cycle (equivalent to a path change of $\lambda/2$) by electronics, thereby providing a least count of $\lambda/2$, $\lambda/4$ or $\lambda/8$ respectively. The pulse count is converted to displacement value with a built-in scale factor.

In one of the commercial version of the laser interferometers, two slightly different frequencies of the laser are used. These two frequencies are obtained by Zeeman splittling, and an heterodyne detection is used for the measurement. The so called a.c. interferometer is more reliable and allows velocity measurement in addition to displacement.

9.4   Diffraction strain gauge

The gauge makes use of the phenomenon of diffraction for the measurement of very small displacements, and hence the strain. A schematic

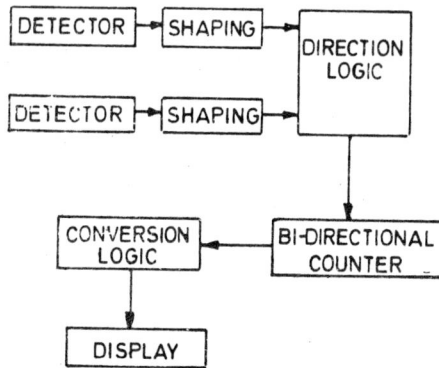

Fig. 9.3 Block diagram of signal processer

of this is shown in Fig. 9.4. A slit with independent jaws is cemented to the test member such that its jaws are parallel. The distance between the fixed points of each jaw of the slit on the test member is the gauge length. Let it be denoted by L. The laser beam is incident normally on the slit. The slit diffracts the beam into various orders which are observed on the screen distant D from the slit. Let $2b$ be the slit width. The irradiance distribution in the far field diffraction pattern of the slit is given by

$$I(p) = I_0 \left(\frac{\sin(kpb)}{kpb}\right)^2,$$

where p refers to the sine of the angle which the direction p makes with the central direction ($p = 0$). The irradiance is zero when $p = \pm \frac{m\lambda}{2b}$; m is an integer. If the cartesian coordinate at the observation plane along a direction perpendicular to the slit is represented by x, then the zero's of the irradiance distribution occur when

$$x_m = \pm m \frac{D\lambda}{2b}.$$

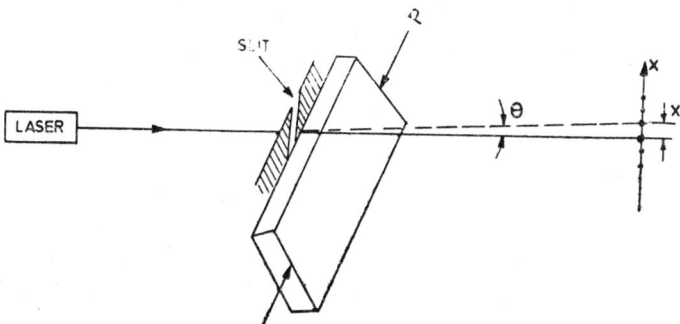

Fig. 9.4 Diffraction strain gauge

The secondary maxima are equidistant with a spacing of $D\lambda/2b$. The distance between the first zero's bounding the central maximum is $D\lambda/2b$.

Any change in the slit width due to loading will result in the corresponding change in the diffraction pattern. A tensile strain will contract the pattern, and the compressive strain will elongate it. The separation between the zero's of the diffraction pattern of the slit at the observation plane, in the absence of loading, is given by

$$\bar{x}_1 = \frac{D\lambda}{2b}.$$

When the specimen is loaded, the slit width changes. Let the slit width on loading be $2b'$. The slit width $2b'$ is related to the slit width $2b$ according to the relation

$$2b' = 2b \pm \epsilon L$$

where ϵ is the strain; positive sign refers to the tensile strain. The separation between the zero's of the irradiance distribution is now given by

$$\bar{x}_2 = \frac{D}{2b + \epsilon L}.$$

From these equations, the value of strain ϵ is obtained as

$$\epsilon = \frac{2b}{L}\left(\frac{x_1}{x_2} - 1\right).$$

For tensile strain $\bar{x}_1 > \bar{x}_2$. Therefore in practice, one measures the separations of zero's in the pattern before and after loading. $2b/L$ is taken as a parameter of the gauge. A smaller value of $2b/L$ indicates higher sensitivity. The slit width $2b$ initially is so chosen that the zero's in the diffraction pattern are reasonably well separated.

The cementing of the slit to the test member may not always be possible. Even if it is possible, if requires great ingenuity on the part of the experimenter to cement the jaws parallel to each other. The measurement of the distance between the fixed points may also be equally frustrating. But in actual practice it is possible to have gauge length smaller than a millimeter. It may also not be possible to measure steep slope gradients with this method as the slit jaws may not remain parallel on loading.

A variant of this method is to draw two fine lines parallel to each other at the area of interest on the test member. The separation between the lines is the gauge length. An unexpanded laser beam incident on the pair of lines is diffracted, and an interference pattern between the two diffracted waves in reflection is observed on the screen. Spacing of the interference fringes whose orientation is parallel to the engraved lines depends on their separation. If the separation between the two lines is L and the interference pattern is observed at a distance D from the test member, the fringe spacing $\bar{x}_1$ is given by

$$\bar{x}_1 = \frac{D\lambda}{L}.$$

On loading, the separation between the two lines changes, and hence the fringe width. The fringe width $\bar{x}_2$ obtained after loading, is given by

$$\bar{x}_2 = \frac{D\lambda}{L'} = \frac{D\lambda}{L(1 \pm \epsilon)}.$$

From these equations, the strain ϵ is given by

$$\epsilon = \left(\frac{\bar{x}_1}{\bar{x}_2} - 1\right).$$

The fringe width can be measured with micrometer eye-piece, when laser of relatively low power output is used. The method requires accurate measurement of the fringe width. It however does not require an optically polished and flat surface. Of course the surface roughness degrades the fringe contrast.

9.5 Laser doppler anemometer (LDA)

Laser anemometer enables the measurement of instantaneous velocity of a gas or a liquid flowing in a glass walled channel. The unique features of this instruments are:
  (i)   non-contact masurement
  (ii) excellent spatial resolution
  (iii) very fast response to fluctuating velocities
  (iv) no transfer function involvement, the output voltage is linearly proportional to the velocity, and
  (v)   measurement possibilities in both gaseous and liquid flows.
These features establish the superiority of Laser anemometer over hot-wire anemometer. It is, however, restricted in usage to bubbly flows; the flows that carry suspensions to scatter light. The particle density of suspensions should not be less than 10^{10} particles/mm^3. The suspension size in gases ranges from 1 to 5 μm and in water from 2 to 10 μm. The suspension particles should follow the flow faithfully.

The theory of LDA is based on the principle of Doppler shift. The frequency of radiation scattered by a particle moving relative to a radiating source is changed by an amount that depends on the velocity and the scattering geometry. Consider a laser beam of circular frequency ω_0 propagating along the unit vector $\vec{C}_1$. If the light scattered along the unit vector $\vec{C_2}$ by a **scatterer** moving with velocity V is examined, its circular frequency ω_s is given by

$$\omega_s = \omega_0 + n_0\, \vec{V} \cdot (\vec{K_s} - \vec{K_l})$$

where n_0 is the refractive index of the medium and $\vec{K_l} = \dfrac{2\pi}{\lambda}\,\vec{C_l}$ are the propagation vectors. The Doppler frequency $\Delta\nu_D$ $[= (\omega_s - \omega_0)/2\pi]$ is given by

$$\Delta\nu_D = \frac{2n_0\, V}{\lambda_0}\, \sin(\theta/2)\, \sin\beta$$

where angles θ and β are shown in Fig. 9.5.

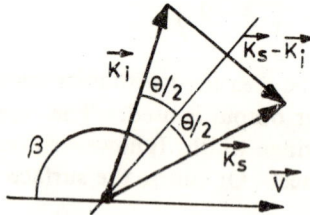

Fig. 9.5   Geometry for scattering

If $\beta = \pi/2$, then the velocity V is given by

$$V = \Delta\nu_D \cdot \frac{\lambda_0}{n_0}\, \frac{1}{2\sin(\theta/2)}$$

The velocity of scatterer is linearly related to the Doppler frequency. This is the basic formula for the laser anemometry.

There are two distinct approaches in vogue to explain the working of an LDA. These are:

(i) Reference beam mode, and
(ii) Interference fringe mode.

In the reference beam mode, laser light is split into two beams which are directed at the point of measurement in the field at an angle θ. The light scattered in the direction of reference beam is picked up and photo-mixed in the photo-multiplier tube with the reference beam propagating in the same direction as shown in Fig. 9.6(a). The photo multiplier tube (PMT) yields a signal at the Doppler frequency. The reference beam need not traverse through the same volume, but can be added to the scattered beam at PMT. The reference beam strength is considerably smaller than that of the other beam, so that scattered beam is relatively stronger and a good heterodyne signal from PMT is obtained. This mode of operation can be used with advantage when the suspension density is fairly high.

Further the alignment tolerance required for the heterodyne process is not so critical. Thus, this mode of operation is relatively easy to use.

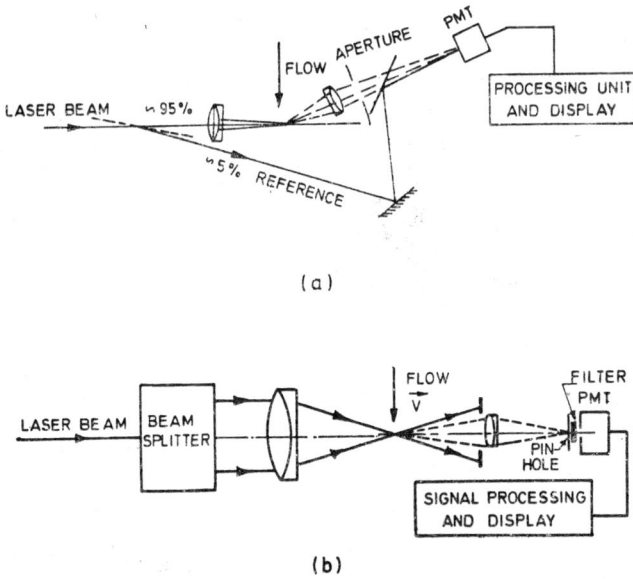

(a)

(b)

Fig. 9.6   Schematic of LDA (a) Reference mode, (b) Fringe mode

Working principle of the interference fringe mode gives a better insight of the appearance of Doppler signal. The schematic of fringe mode anemometer is shown in Fig. 9.6(b). The beam from the laser is symmetrically split by a beam splitter and then focussed in the flow by a lens L_1. In the region of intersection of these two beams, Young's interference fringes are formed with a spacing $\bar{x}$ given by

$$\bar{x} = \frac{\lambda f}{s} = \frac{\lambda_0}{2n_0 \sin (\theta/2)}$$

where s is the beam separation and f is the focal length of lens L_1. The fringes are parallel to the bisector of angle between the two beams. As a particle moves across this fringe pattern, it passes through planes of maximum and minimum intensities. When it is at the bright fringe (plane of maximum intensity) it scatters more light. Therefore as the particle traverses the interference field normal to the fringes, it scatters the light which generates a periodic signal with a frequency

$$2\left(\left(\sin \frac{\theta}{2}\right)\frac{n_0}{\lambda_0} V\right).$$

The scattered light is collected by another lens in the pick-up unit. The sample volume (interference region) is imaged on a pin-hole in front of the detector. This is the most critical alignment to get a good Doppler signal. A filter for He–Ne laser is also mounted behind the pin-hole so that the anemometer can be run under normal lighting conditions. The output of the photo-detector when displayed over CRO appears as shown in Fig. 9.7(a). The velocity signal rides over a strong d.c. background. At this point, there are two approaches available for signal processing. One can either work in time-domain and take the auto-correlation of the photo–current. For this the photo-current is fed to an autocorrelator. The fourier transform of the auto-correlation function is the power spectrum. The output of the auto-correlator can be inter-faced to a computer to give the peak frequency of the signal and hence the velocity. In the other approach, the signal can be directly fed to a spectrum analyser which yields the full spectrum [Fig. 9.7(b)].

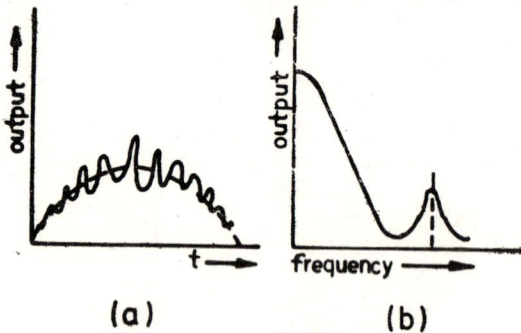

Fig. 9.7 A typical signal output
(a) Time domain
(b) Frequency domain

The interference planes do not exist over all the volume but are confined to a region where the two beams are visually seen to intersect. The finite region over which the interference fringes are observed may be called the *sample volume*. In fact the actual sample volume is much smaller than this. If the irradiance distribution at the lens surface is expressed by $I(r) = I_0 \exp(-r^2/2\sigma^2)$, the irradiance distribution at the focus will also be gaussian and have the form $\exp(-r'^2/2\sigma'^2)$, where 4σ, and $4\sigma'$ are the beam width at the lens surface and the focus respectively. It can be shown that $\sigma' = \lambda f/4\pi n_0\sigma$.

The sample volume is shown in Fig. 9.8, and can be approximated by a body of rotation whose cross-section is a parallelogram. Further, it may also be noted that the minimum beam waist would not lie at the focal plane but for systems with $f/D > 10$, the minimum beam waist can be taken to lie at the focal plane where D is the aperture of the lens. The quality of lenses used in the anemometer need not be very good. Plano-convex lenses are found to serve well. The flow to be measured should be accessible and hence optical windows are invariably installed for the passage of beams. It can be shown that the presence of windows only displaces the location of the sample volume, and does not influence the Doppler frequency in any way.

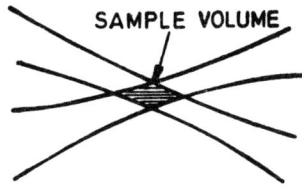

Fig. 9.8   Intersection of two gaussian beams; sample volume

Some situations compel the experimenter to use the back scattered light. The intensity of back scattered light is many orders of magnitude smaller than that of the forward scattered beam. In order to get good signal to noise ratio from the back scattered signal, and to eliminate the d.c. component arising due to reflections from stationary components some dye particles which fluoresce at a different wavelength than that of the laser are used in the flow. The fluorescence radiation from the dye can be conveniently filtered out and directed to the PMT. When both components of the flow velocity are to be monitored, the scattered light is picked up in two perpendicular planes by the two independent photo-detectors. A commercial system using two wavelength argon ion laser is available for this kind of applications. Another unit with flow speed range from 0.3 cm/s to 300 m/s and a good frequency response upto 150 KHz is commercially available. Accuracy of this anemometer is 1% of full scale reading.

9.6   Holography and hologram interferometry for stress-analysis

Necessary preliminary work on holography was done by Gabor in 1947-50. The later researchers refined some of the concepts but nothing extraordinary happened during the next decade. The appearance of laser in 1960 revived the subject and a whole lot of fresh activity started. The recording of 3-D diffused objects using principles borrowed from communication theory by Leith and Upatnieks provided added impetus to the activity. Soon it was realised that the waves scattered from diffused objects can be made to interfere providing information with interferometric accuracy. So a new kind of interferometery now termed as hologram interferometry was born. It became possible to obtain interference between waves obtained from time-delayed events. Applications to a variety of problems was subsequently possible.

The researchers in the area of holography were not very happy with the presence of grannular structure on the object, a physical characteristics of a coherent field, and began to invent ways to reduce/eliminate them. The speckling as it is called today was responsible for ultimate resolution in hologram interferometry. Another group of researchers thought better to live with speckling and tried to exploit this phenomenon.

There thus grew another area now called speckle photography, speckle interferometry etc. This provided a very good method with adjustable sensitivities to the experimental stress analyst. These will be briefly discussed now.

HOLOGRAPHY

Let us consider an object that is illuminated by a spatially coherent wave. Let the wave transmitted/reflected from the object be represented by $\hat{O}$ [$= O_0 (x, y, z) \ e^{i\phi_0(x,y,z)}$]. The information about the amplitude, relative position and direction are contained in $O_0 (x, y, z)$ and phase $\phi_0 (x, y, z)$. If the object wave is to be recorded on a square detector $\left(\text{one that responds to } \left|\hat{O}\right|^2\right)$ some kind of technique that converts phase information to the irradiance variation is to be used. Fortunately, it can be accomplished with the help of interferometry. Therefore, the first step that involves the recording of complete information about the object is to add a coherent reference wave to the object wave at the plane of recording.

Photographic emulsion either on cellulose film or glass plate, is assumed as the recording medium. Further, the photographic plate is assumed to be positioned at $z = 0$ plane, and the object and reference waves at $z = 0$ plane are designated by $\hat{O} (x, y)$ and $\hat{R} (x, y)$ [$= R_0 e^{i\phi_R}$]. The exposure E recorded on the photographic plate during the exposure time T is given by

$$E = T \left|\hat{O} (x, y) + \hat{R} (x, y)\right|^2$$

$$= T \left[\left|\hat{O}\right|^2 + R_0^2 + \hat{O} \hat{R}^* + \hat{O}^* \hat{R}\right]$$

$$= T \left[\left|\hat{O}\right|^2 + R_0^2 + 2 O_0 R_0 \cos (\phi_0 - \phi_R)\right]$$

The phase information ϕ_0 has been now transformed to the exposure variation. A processed recording of E is called a hologram. In a simple language it is a record of an interference pattern. If both the waves were plane waves angularly separated, the hologram will be a grating. Since the object wave is a complex one, a hologram may be termed as a generalised grating. It is, however, obvious that the recording of hologram involves the interference between the object wave and reference wave and thus demands the fulfilment of all those conditions needed for interference. A hologram recording schematic is shown in Fig. 9.9(a).

A hologram can now be assigned an amplitude transmittance $\hat{t}$, in the form

$$\hat{t} = \hat{t}_0' - f(E)$$

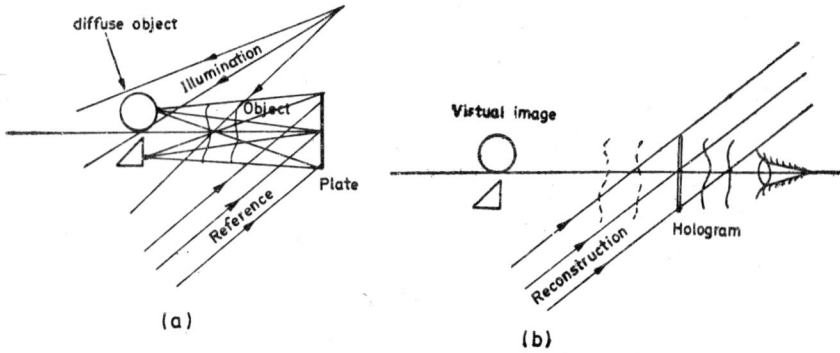

Fig. 9.9 Schematic for (a) Recording and (b) Reconstruction of wave

where $\hat{t_0'}$ is the base transmittance and $f(E)$ is some function of exposure, properties of recording medium and development process. For the sake of simplicity it is assumed that $\hat{t}$ is linearly related to E, i.e.

$$\hat{t} = \hat{t_0'} - \beta T \left\{ \left| \hat{O} \right|^2 + R_0^2 + \hat{O}\overset{*}{\hat{R}} + \hat{O}^*\hat{R} \right\}$$

$$= \hat{t_0} - \beta T \left\{ \left| \hat{O} \right|^2 + \hat{O}\overset{*}{\hat{R}} + \hat{O}^*\hat{R} \right\}$$

where the constant term $\beta T R_0^2$ has been absorbed in $\hat{t_0}$. When a collimated beam of amplitude $\hat{B} \left[= B_0 e^{i\phi_B} \right]$ is incident on the hologram, the amplitude of transmitted wave can be expressed as $\hat{B}\hat{t}$. Therefore, there are four waves present: the first wave $\hat{B}\hat{t_0}$ is a collimated wave propagating in the direction of $\hat{B}$. The wave proportional to $\left| \hat{O} \right|^2$ has double the frequency contents of $\hat{O}$ and is centered about the zero frequency (along $\hat{B}$). The wave proportional to $\hat{B}\hat{O}\overset{*}{\hat{R}}$ gives rise to the virtual image of the object and if $\hat{B} = \hat{R}$, this term is identical to the object wave but for a multiplication factor, and would generate an object image identical to the object in all respects. The wave proportional to $\hat{B}\hat{O}^*\hat{R}$ gives rise to the conjugate object image. Figure 9.9(b) shows the schematic of reconstruction of virtual image. It may again be noted that the reconstruction of the object wave is due to diffraction; diffraction of wave $\hat{B}$ on the generalised grating (hologram). Therefore, the spatial and temporal coherence requirements are not very stringent. A hologram can be reconstructed by a wave of any wavelength but the image would exhibit scaling and wavelength dependent aberrations.

Hologram recording geometries

There are a number of recording geometries with distinct features. However, only two main categories, namely in-line and off-axis are considered here. Most of the remaining configurations can be included in any one of the two but are given distinctive names due to many other parameters.

(a) *In-line geometry*: The object and reference waves propagate along the same direction and have relatively low spread. Looking at the tramsmitted fields from the hologram, one can easily say that all the four fields propagate along the same direction. The virtual and conjugate images are formed equidistant on either side of the hologram. Looking through the hologram, if one focusses at the virtual image, the conjugate image appears as defocussed image. Further, if $|\hat{O}|^2 \geqslant |\hat{R}|^2$ then the term $|\hat{O}|^2$ will be very dominant and the object images $\hat{O}$ and $\hat{O}^*$ will be completely submerged in the noise due to $|\hat{O}|^2$. For this reason, it is limited to a distinctly different class of objects.

(b) *Off-axis geometry*: The reference wave is added at the photographic plate at an angle with respect to the object wave. In reconstruction, the virtual and conjugate images are angularly separated with respect to the direct beam. However, in order to avoid overlap between $|\hat{O}|^2$ and $\hat{O}$ terms, the angle between the reference and object beam must be greater than a certain minimum determined from the object spectrum. Thus both the disadvantages of in-line holography have been eliminated in off-axis holography. It however puts a severe demand on the resolution of the photographic plate.

Requirements for holography

1. Recording of off-axis hologram requires special emulsions with resolution in excess of 2000 l/mm. These emulsions tend to be slow and thus require longer exposure time.

2. Due to long exposure needed to record holograms, a stable laser and vibration isolated system are required. During the course of exposure, no component in the setup should move to an extent that may cause path difference changes of more than $\lambda/8$ for good recording.

3. In order that $\hat{i}$ and E remain linearly related, the ratio of irradiances of reference to object waves should lie between 3 and 10.

4. Photographic processing requires controlled temperatures for various baths.

HOLOGRAM INTERFEROMETRY

It has been shown that an object wave can be stored in a hologram and released at a later time when required. This capability of hologram has given rise to a new branch of interferometry called hologram interferometry. There are two ways in which it can be used to compare two states of an object. They are

(a) Real time/single exposure/live-fringe hologram interferometry.

(b) Lapsed time/double exposure/frozen fringe hologram interferometry.

Real-time HI

A holographic record of an object wave is made using one of the off-axis configurations. After processing, the hologram is replaced accurately at the position it occupied during recording. On interrogation with the reference wave, a wave indentical to the object wave at photographic plate during recording but with a phase change of π due to photographic processing will be released. In addition to this the object wave from the object will pass through the hologram with little attenuation. The object waves transmitted through the hologram, as also released from it would interfere destructively. The object would therefore appear dark. In practice there are always a few residual fringes due to the wet photographic process. If the object is now subjected to some kind of deformation, so that only the phase of the object wave is changed interference fringes would be observed in real time.

The technique has its own merits and demerits. It is used to avoid sign ambiguity for displacement. The fringe contrast is relatively poor but can be improved by exercising control over the intensity of the reference wave. Although it is possible to continuously monitor the state of the object till the resolution limit, its main disadvantage is that the technique cannot be used if any of the optical elements in the setup is disturbed accidentally or one wishes to evaluate the hologram leisurely.

Double exposure HI

Two object waves belonging to two different states of the object are recorded sequentially on the photographic plate. On reconstruction, these two waves are released from the hologram. These waves interfere and give information about the change the object has undregone in between the exposures. Since the diffraction efficiency is equally shared, the fringes of almost unit contrast are formed. Further if a plane wave or a spherical reference wave is used for recording, the hologram can be analysed leisurely at a latter time. The main disadvantage of the double exposure HI is that it records only two states of the object and hence continuous monitoring is not possible. However a full range can be covered by making a large number of double exposure holograms.

Both live-fringe HI and double exposure HI can be used for vibration analysis. It is also possible to record many cycles of the vibrating object continuously on the photographic plate. Reconstruction of holo-gram provides information about the vibration modes and their ampli-tudes. This is called time-averaged HI. Recently sandwich HI has been developed. It finds applications in solving many complex problems and offers certian advantages over double exposure HI.

Reduction of hologram data: It has been assumed that the deformation of the object changes only the phase of the object wave. If the point (x, y, z) on object moves to another point (x', y', z') with the vector displacement of $\vec{L}$ on loading, it can be shown that the phase difference is given by $(\vec{K_1} - \vec{K_2}) \cdot \vec{L}$ where $\vec{K_1}$ and $\vec{K_2}$ are the wave vectors of illuminating and scattered waves. Thus for the mth fringe to form

$$(\vec{K_1} - \vec{K_2}) \cdot \vec{L} = 2\pi m : m = 0, \pm 1 + 2 \ldots$$

Neither $\vec{L}$ nor m are known. Therefore, if one changes the direction of observation but still looks at the point of interest, a number of fringes will be found to cross the point of observation. In fact one can set up the following equations

$$(\vec{K_1} - \vec{K_{21}}) \cdot \vec{L} = 2\pi m_1,$$

$$(\vec{K_1} - \vec{K_{22}}) \cdot \vec{L} = 2\pi m_2,$$

$$(\vec{K_1} - \vec{K_{23}}) \cdot \vec{L} = 2\pi m_3,$$

where $\vec{K_{ij}}$ are the directions of observation and hence are known. Another set of three equation can now be setup in terms of the number of fringes passed through the point when the direction of observation is changed. These equations are:

$$(\vec{K_{22}} - \vec{K_{21}}) \cdot \vec{L} = 2\pi (m_2 - m_1).$$

$$(\vec{K_{23}} - \vec{K_{22}}) \cdot \vec{L} = 2\pi (m_3 - m_2),$$

and

$$(\vec{K_{21}} - \vec{K_{23}}) \cdot \vec{L} = 2\pi (m_1 - m_3).$$

A set of equation of this type is used to find the components of $\vec{L}$. There are a number of other approaches to evaluate an interferogram.

HOLOGRAPHIC FLOW VISUALISATION

Shadowgraphy, schlieren photography and optical interferometry have been used for flow visualisation. The optical interferometry provides refractive index distribution and Mach-Zehnder interferometer is most commonly used for this purpose. Because of its simplicity and state of development, the $M - Z$ interferometer is likely to remain the most common interferometer for analysis of compressible flows. The unique features of hologram interferometry, however, make possible new types of measurements and extend the range of optical and environmental conditions in which interferograms can be recorded.

The holography can be applied to aerodynamic problems in two distinctly different ways. Either (1) a single exposure or (2) double exposure holography could be used; the latter gives an interferogram for analysis.

The experimental arrangement for recording a hologram is similar to that described earlier, the object beam traverses through the object. Usually a diffuse illumination is used. Using a pulse laser, say a ruby laser, a hologram of an event in the flow field is recorded. If the pulse duration is smaller than the duration of the transient event of interest, the event may be assumed to be frozen during recording. This single-exposure hologram is a permanent record of the optical wave that has traversed the flow field. The hologram can be interrogated with a reference beam, thus releasing the object wave which can be studied leisurely using shadowgraph or Schlieren methods. If a He-Ne laser is used for reconstruction of the object wave from the hologram, care should be taken to include the effect of wavelength change while conducting the analysis on the processed image. It is possible to make two records on the photographic plate, one without the flow and the second with the flow on. On reconstruction, the image of the flow field carries information about it in the form of interference fringes. This is the standard double exposure hologram interferometry. Some lasers are equipped with a provision to give two pulses in succession with a time delay that can be varied by the experimenter. Therefore, the difference between two transient events can be recorded for analysis at leisure. Some of these interferograms are awfully good.

The use of hologram interferometry does not require high quality optics, as the phase errors cancel out during reconstruction. Besides it is relatively simple to perform experimentally.

HOLO-PHOTO-ELASTICITY

It has been shown that photo-elasticity gives information about the difference of principal stresses $(S_1 - S_2)$ and their directions. In order to find out the magnitude of S_1 and S_2 separately, either $(S_1 + S_2)$, or S_1 or S_2 must be known. Interferometry is used to get the sum of stresses. If holography is applied to photo-elasticity, it is posible to obtain interference patterns belonging to $(S_1 - S_2)$ and $(S_1 + S_2)$ respectively. Because of this interesting feature, hologram interferometry has been employed to study photoelastic models by a number of researchers. A new name, holo-photoelasticity, is given to this activity. A schematic of an interferometer suitable for photo-elastic work is given in Fig. 9.10. The specimen is transilluminated. The reference beam is a circularly polarised plane wave. In one of the methods applied to photo-elasticity, two hologram are made:

   (i) a single exposure hologram, after the object (model) is loaded, and

   (ii) a double exposure hologram, first exposure before loading and second exposure after loading the object.

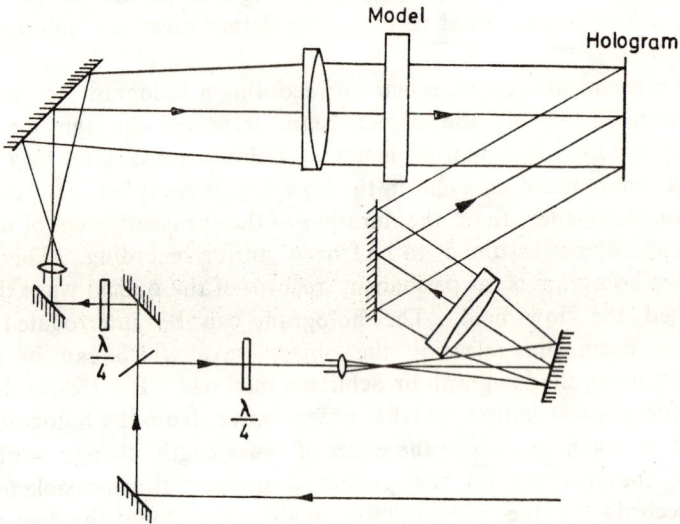

Fig. 9.10   Holo-photo elastic interferometer

The first hologram on reconstruction gives isochromatics alone, while the second hologram gives both isochromatics and isopachics. It would become obvious if a theoretical base is provided to the analysis.

When a model is loaded, it becomes birefringent; the directions of principal refractive indices being parallel to the directions of principal stresses. The principal refractive indices are related to the principal stresses by Maxwell's equations as

$$n_1 - n_0 = AS_1 + BS_2,$$

and

$$n_2 - n_0 = BS_1 + AS_2,$$

where n_0 is the refractive index of the unstressed model, n_1, n_2 are principal refractive indices of the model and A, B are the stress-optic coefficients to be determined by calibration. On combining these equations one gets

$$n_1 - n_2 = (A - B)(S_1 - S_2).$$

This is called the stress-optic law.

A wave incident on a birefringent model will decompose into two orthogonally polarised waves which propagate the model with different velocities and hence have a path difference Δ given by

$$\Delta = (n_1 - n_2) \, d,$$

where d is the thickness of the model. Therefore the emergent wave from the model is in general elliptically polarised.

When a single-exposure hologram of a stressed model is made, one essentially records interference patterns between a circularly polarised reference wave and orthogonally linear polarised beams with a path difference Δ generated in the model. On reconstruction, an interference pattern superposed on the reconstructed image is observed. The irradiance distribution in the interference pattern is given by

$$I = I_0 \cos^2 \left[\frac{\pi}{\lambda} (A - B)(S_1 - S_2) \, d \right].$$

This gives isochromatics whose positions are given by

$$\frac{\pi}{\lambda} (A - B)(S_1 - S_2) \, d = m\pi : \text{(bright isochromatics)}$$

$$\text{or } (S_1 - S_2) = \frac{m\lambda}{(A - B) \, d}.$$

The fringes are therefore loci of constant $(S_1 - S_2)$. The dark isochromatics are governed by the condition

$$\frac{\pi}{\lambda} (A - B)(S_1 - S_2) \, d = (2m + 1) \, \pi/2.$$

When double exposure (first exposure before loading and second after loading) hologram is recorded, reconstruction from the hologram results in an image overlaid with a pattern whose irradiance distribution is of the form

$$I = I_0 \left[3 + 4 \cos \left\{ \frac{\pi}{\lambda} (A + B)(S_1 + S_2) \, d \right\} \right.$$
$$\left. \cos \left\{ \frac{\pi}{\lambda} (A - B)(S_1 - S_2) \, d \right\} + \cos^2 \left\{ \frac{\pi}{\lambda} (A - B)(S_1 - S_2) \, d \right\} \right]$$

This intensity distribution contains information about both the sum $(S_1 + S_2)$ and difference $(S_1 - S_2)$ of the principal stresses. The fringe contours of the same value of $(S_1 + S_2)$ are called isopachics. If the spacings of isochromatics and isopachics are sufficiently different, and if they are nearly orthogonel, they are easily recognisable; the isochromatics are recognised by their half-tone grey fringes while the isopachics are dark. Further when an isopachic fringe crosses an isochromatics, its irradiance changes from bright to dark or vice-versa, so the isopachics are easily recognised. In the method discussed here, iscochromatics are separately obtained and hence isopachics can be easily found from the combined pattern.   The constants A and B can be obtained by any one of the calibration tests on the model material: (i) tensile test, or (ii) compression test, or (iii) bending test.   With the knowledge of A, B and m, the stresses S_1 and S_2 can be found.   It may be mentioned here that isoclinics can also be obtained on the same setup with minor modifications.

9.7 Speckle photography

Annoying grannular appearance observed when the object is illuminated by a coherent light has been used for measurement purposes.   It has been utilised to obtain in-plane displacement for which hologram interferometry is least suited.   Also out of plane displacement, tilt, and slope contours have been measured with speckle photography.   The technique provides a sensitivity in between that of interferometry and Moiré method; often the sensitivity can be continuously varied in the filtering process, even after the conduct of experiment.   Requirements for mechanical stability are much less sever than those for hologram interferometry.

Consider two different arrangements [Figs. 9.11 (a) and (b)] for observing speckles: (a) A diffuser of size d is illuminated by a laser light, the field at a distance z from the object will have speckled appearance with an average speckle size of $\lambda z/d$.   This is called objective speckle field. (b) A lens of aperture D and focal length f makes an image of an object. The image is speckled and the average speckle size in the image is $\lambda v/D$, where v is the image distance.   This is the subjective speckle pattern.

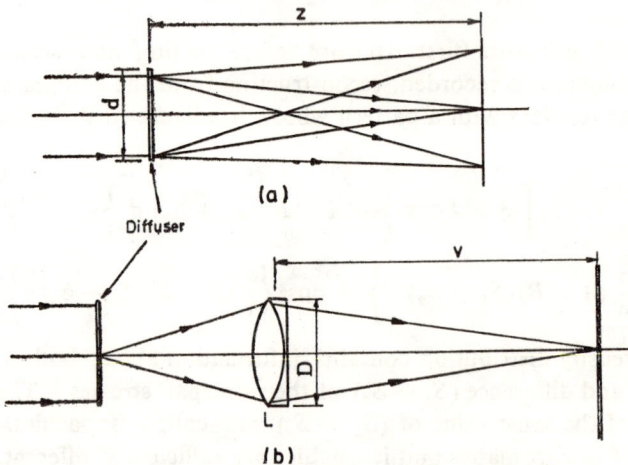

Fig. 9.11   Speckle pattern formed
(a)   In fraunhoffer plane (Objective speckle pattern),
(b)   In image plane (Subjective speckle pattern)

When the object undergoes displacement, the speckle field also shifts.   In order to understand the basic concepts of speckle photography as applied to the measurement, consider the following two experiments. The objective of the first experiment is to measure in-plane displacement, while that of the second is to measure the tilt of a diffuse object.

IN-PLANE DISPLACEMENT MEASUREMENT

Figure 9.11(b) shows a schematic for the measurement of in-plane displacement. A lens L makes an image of a diffuse object on the recording plane. The object is illuminated by laser light. At the image plane speckles of an average size of $\lambda v/D$ are observed. This speckle field is recorded on a high resolution photographic plate. When the object undergoes an in-plane displacement, say, by L_0, the speckle pattern will also translate on the recording plane by ML_0, where M is the magnification of the system. The shift of speckle pattern is independent of the angle of illumination θ when a diffuse opaque object is illuminated by a collimated beam. The shifted speckle pattern is also recorded on the same photographic plate. The photographic plate—on development called a specklegram—contains two indentical speckle patterns which are shifted by an amount ML_0. In order to find the magnitude and direction of the displacement, the specklegram is illuminated by a narrow beam of laser light. A set of Young's fringes whose orientation is perpendicular to the direction of displacement is observed in the speckle halo. Their fringe width is given by

$$\bar{x} = \frac{\lambda z}{ML_0}.$$

The observation of fringe width $\bar{x}$ is made at a distance z from the specklegram. Therefore

$$L_0 = \frac{\lambda z}{M\bar{x}}.$$

For a uniform translation of the whole object, the fringe width and their orientation must remain the same when the specklegram is scanned at its various regions. For non-uniform displacement, the specklegram is scanned and the fringe widths and orientations are recorded. Alternately a filtering arrangement that provides variable sensitivity can be used to obtain full field displacement pattern.

MEASUREMENT OF TILT

A schematic of the experimental setup for the measurement of tilt is shown in Fig. 9.12. The speckles are recorded at the back focal plane of the lens L where the average size of the speckles is $\lambda f/D$. The speckle is formed by light travelling in a particular direction. The first exposure record is the initial state of the object as a speckle field. When the object is tilted by an angle a, there is a relative phase change $\Delta\phi$ which is given by

$$\Delta\phi = \frac{2\pi}{\lambda}\left(1 + \cos\theta\right) a\, x,$$

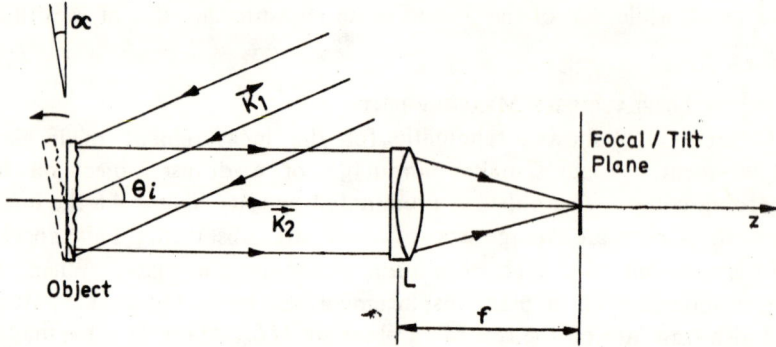

Fig. 9.12 Schematic for the measurement of tilt

The speckles are therefore translated by an amount L_f in the back focal where L_f is given by

$$L_f = (1 + \cos \theta) \, a \, f$$

The second exposure, therefore, records a similar but shifted speckle pattern on the same photographic plate. If the specklegram is scanned by a laser beam, Young's fringes with spacing $\bar{x}$ are observed at a distance z from it where $\bar{x} = \dfrac{\lambda z}{f(1 + \cos \theta) \, a}$. From the geometry and the measurement of the fringe spacing, tilt angle can be easily measured.

These two simple experiments are indicative of a large number of possibilities that speckle photography offers. The main advantages of speckle photography over hologram interferometry are its simplicity, and minimal requirements on mechanical stability.

9.8 Temperature sensor

A number of methods have been discussed to measure temperature in chapter 8. An interferometer can be used to measure temperature by measuring linear expansion of a suitable material. It is possible to discriminate path changes of the order of $\lambda/20$. Therefore using a brass rod of 1 cm length it is possible to measure temperature difference of 0.1°C. This is an extremely poor sensitivity expected from an interferometric measurements. The situation changes dramatically if the temperature sensing element is a part of an active interferometer. A small change in the cavity length results in large variation in frequency according to the relation $\dfrac{\Delta \nu}{\nu} = \dfrac{\Delta L}{L}$. Again using the same date, $\Delta L = 34.2 \times 10^{-6}$ cm/cm °C, the frequency change of He-Ne laser will be $\Delta \nu = 16.2\,\mathrm{MHz}$. A one degree celsius change in temperature would result in a frequency change of 16.2 MHz. Frequency differences of the order of

1 KHz or even lower can be measured accurately, thereby measuring temperature variations of the order of 10^{-4}°C. Figure 9.13 shows a schematic of an arrangement to measure small temperature difference. Temperature sensing etalon is inserted in the cavity. The frequency of the laser will shift as the length of the etalon changes due to temperature. The frequency shift is measured by heterodyning the laser beam with another beam from a laser which is frequency stabilised.

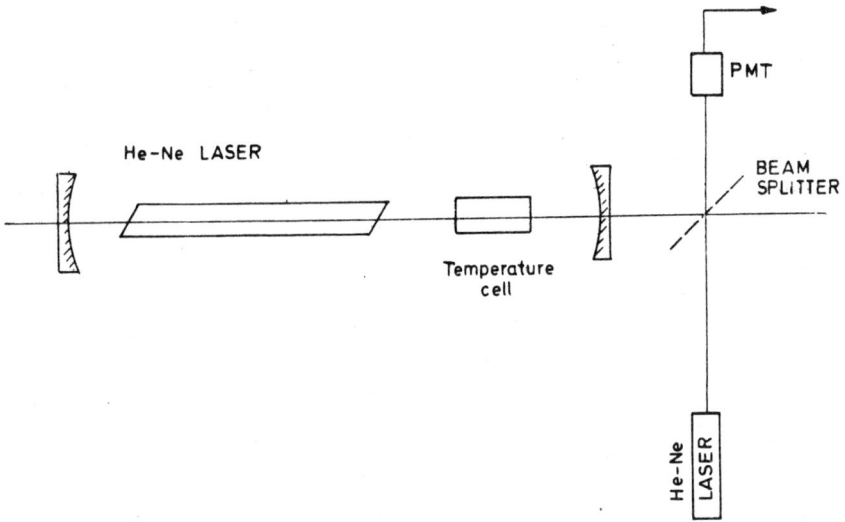

Fig. 9.13   Measurement of temperature using heterodyne principle

9.9   Seismographic measurement

An extremely sensitive arrangement for displacement, and consequently to detect seismographic events is shown in Fig. 9.14. The instrument possesses a sensitivity that is atleast ten times better than that of the

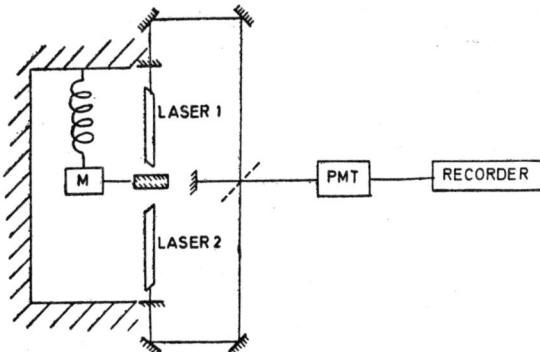

Fig. 9.14   Schematic of a laser seismograph

present day seismometers and can detect underground nuclear blasts and earthquakes. It has a very good dynamic response.

One of the mirrors of the gas laser is attached to a large suspended mass. When a disturbance in the earth occurs the mass M moves and differentially changes the frequency of the two lasers. The output beams are combined and photo-mixed in a PMT and recorded. The beat frequency $\Delta\nu$ is directly related to the severity of the disturbance. The use of two gas lasers in this fashion completely eliminates the requirement for having a separate frequency stabilised laser for heterodyning.

The applications discussed in this chapter do not exhaust the list of applications of optical methods in the measurement area but are only representative.

Exercises

1. A slit of 200 μm is illuminated by a He–Ne laser light ($\lambda = 0.6328$ μm) and the diffraction pattern is observed at a distance of 150 cm. On loading the specimen the diffraction pattern expands by 10%. Calculate the slit width and hence the strain.
2. Calculate the flow of water ($n = 1.33$) in m/sec, when a beam from a He–Ne laser ($\lambda = 632$ nm) scattered at an angle of 2.7° beats with a reference beam to give a Doppler frequency of 10 MHz.
3. A beam of 10 mm diameter is separated by a distance of 30 mm. The beams are brought to focus by a lens of 100 mm focal length. Calculate (i) the beam waist size and hence the sample volume at the focal plane, and (ii) the flow velocity normal to the fringes when a signal at 150 KHz is observed.
4. A diffuse object is translated in between two speckle records by an amount of 10 μm. The Young's fringes are being observed at a distance of 100 mm. Calculate the fringe width.
5. Discuss the influence of linearly polarised reference beam on the quality of hologram in the study of holo–photo–elasticity.
6. Compare the holographic method of stress analysis with the photo–elastic method and discuss their relevant features.
7. A quasi-plane diffuse object is illuminated by a collimated laser beam which makes an angle of 60° with it. The speckle patterns are recorded at the focal plane of a lens of focal length of 100 mm. In between the recordings, the object is tilted by an angle of 10^{-3} rad. Calculate the spacing of the fringes if the same lens is used for the analysis of specklegram.
8. An etalon has a spacing of 10 cm and the spacer is made of A1. Calculate the beat frequency when the temperature changes by 10^{-3} °C. What would be the corresponding change if the spacer were of quartz? Assume $\lambda = 0.6327$ μm. Also do the calculations for $\lambda = 1.06$ μm.
9. A beam of light is incident on a Wollastan prism. How does the intensity of transmitted beam vary as the prism is rotated when the incident light is (1) linearly polarised, (2) circularly polarised and (3) unpolarised?

Fibre-Optics in Measurement

10.1 Introduction

Fibre optics is finding applications for the measurement of liquid level, fluid flow, temperature, pressure, rotation, magnetic field and electrical current etc. The fibre optic sensors offer many advantages for measurement. The most obvious is the lack of electrical bias at the sensor, thereby easing EMI (electromagnetic interference) problems and providing numerous safety considerations in hazardous areas. The fibre sensor is a low weight transducer, and hence can respond over wide frequency range. It is chemically inert and offers higher accuracy of measurements in most of the cases as compared to the conventional methods.

10.2 What is Fibre Optics?

A fibre is a thin strand of a dielectric material that can conduct light. It usually consists of a cylindrical core of glass, quartz, plastics etc., surrounded by a cladding material. The core material is extremely pure, i.e. the attenuation is very low. A light ray entering the fibre is trapped by total reflection at the core-cladding boundary.

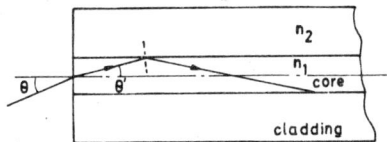

Fig. 10.1 Ray propagation in a fibre.

The ray propagation in a fibre can be understood following Fig. 10.1. The fibre consists of a core of refractive index n_1 surrounded by a cladding of refractive index n_2. The index n_2 is smaller than n_1 so that the conditions for total reflection at the core-cladding boundary may be satisfied. A ray of light is incident at an angle θ at the entrance face. The angle of incidence at the boundary is $(\pi/2 - \theta')$ where θ' is the angle of refraction in the core. If the angle of incidence $(\pi/2-\theta')$ is greater than the critical angle θ_c $(=\sin^{-1} n_2/n_1)$ the ray will be totally reflected. Since the diameter of core is uniform, the angle of incidence will remain unchanged and hence always greater than critical angle. Such a ray will propagate into the core with a little attenuation. The rays in the neighbourhood of optical axis satisfy this condition. As the angle increases, the angle of incidence at the boundary decreases. There is an angle θ_a for which the angle of incidence is equal to the critical angle. Beyond this value of θ the ray suffers both reflection and refraction at each encounter with the interface. This ray will therefore die quickly. Therefore the fibre has an angle of acceptance beyond which it does not conduct the beam. The angle of acceptance θ_a is calculated using the relation

$$\theta_c = \pi/2 - \theta' = \sin^{-1}(n_2/n_1) \tag{10.1}$$

where θ' is given by:

$$\sin\theta_a = n_1 \sin\theta' \tag{10.2}$$

Using Eqs. (10.1) and (10.2) we obtain

$$\sin\theta_a = (n_1^2 - n_2^2)^{1/2} \tag{10.3}$$

The numerical aperture (NA) of the fibre is taken as $n_a \sin\theta_a$ where n_a is the refractive index of the medium in front of the entrance face. It is taken unity for air. The numerical aperture NA is then given by

$$NA = (n_1^2 - n_2^2)^{1/2} \tag{10.3a}$$

Usually n_1 and n_2 are nearly equal. Thus we can express NA as

$$NA = (2n\,\triangle n)^{1/2} = n\sqrt{2\triangle} \tag{10.3b}$$

where $n = n_1 + n_2$, $\triangle n = n_1 - n_2$ and $\triangle = \triangle n/n$.
For a fibre of refractive indices of core and cladding of 1.515 and 1.5 respectively, the numerical aperture is equal to 0.21 which corresponds to $\theta_a = 12.2°$.

The simple approach suggests that the light is conducted by total internal reflections at the boundary inside the fibre. The fibre has a finite NA determined by the refractive indices of core and cladding.

10.3   Fibre Types

The fibre shown in Fig. 10.1 and also in Fig. 10.2(a) is a step index fibre in which the refractive index abruptly changes at the boundary. The refractive index is constant in the core and changes to a slightly lower but constant value in the cladding. The core diameter 'a' in most of the optical fibres is in the range of 4 to 400 μm but the standard is taken as 50 μm. The ray propagates in the fibre in straight lines and is totally reflected at the core-cladding boundary. Obviously light rays propagate through different paths, referred to as different modes. Actually the propagation in thin fibres is studied by Maxwell's equations with appropriate boundary conditions. Consequently there are specific "allowed" ray directions which correspond to setting up electric fields across the core to satisfy the boun-

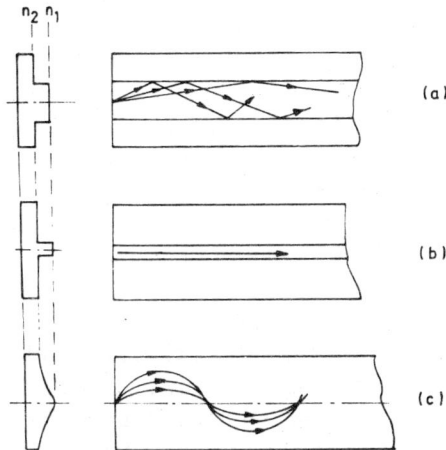

Fig. 10.2 Fibre types (a) Multimode step index.
(b) Single mode step index.
(c) Multimode graded index.

dary conditions. These are the different modes mentioned earlier. Consider a rectangular wave guide (fibre of rectangular cross-section) of cross-section $a \times b$. The boundary conditions that demand zero electric field at the core-cladding interface allow the ray directions that satisfy the condition

$$a \sin \theta' = m \lambda/2 \qquad (10.4)$$

where m is the mode number of the propagating modes; the second mode number is determined by specifying similar boundary conditions in the orthogonal direction. The rays propagating at small angles with the axis will have smaller mode numbers. Since all the rays till the condition for total internal reflection remains satisfied, are allowed, the maximum mode number m_{max} can be expressed as

$$m_{max} = (2a/\lambda)[1-(n_2/n_1)^2]^{1/2} \qquad (10.5)$$

where λ is the wavelength in the waveguide.

The maximum mode number m_{max} can be expressed in terms of the numerical aperture of the fibre as

$$m_{max} = (2a/\lambda)(NA)/n_1 = (2a/\lambda_0)(NA) \qquad (10.6)$$

where λ_0 is the free space wavelength. If m_{max} is less than 2, the fibre is called a single mode fibre. A fibre is of a circular cross-section, and the boundary conditions at the core-cladding interface involve the finite field. A circular fibre of diameter 'a' will support a single mode when

$$\pi a \leqslant 2.405 \; \lambda/\text{NA} \qquad (10.7)$$

Such a fibre is called the single mode fibre and has a core diameter of approximately 5 μm. This is shown in Fig. 10.2(b). Due to extremely small core diameter, it is difficult to launch the light into the fibre; fibre to fibre connection is highly precise and splicing requires greater accuracy. The single mode fibre offers the advantage of no intermodal dispersion as it supports only a single mode.

Another kind of fibre that has low dispersion is multimode graded index fibre shown in Fig. 2.10(c). The typical diameters of core and cladding are also 50 μm aud 125 μm respectively. The refractive index in the core is not constant but varies approximately parabolically. The inhomogeneous core implies that the rays do not travel in straight lines; a ray is continually bent towards the axis and is totally reflected even before it strikes the core-cladding boundary. The refractive index profile is such that the various modes or rays have equal optical path lengths thereby significantly reducing the intermodal dispersion.

10.4 Properties of Optical Fibres

The properties of optical fibres may be divided into two groups: (i) optical, dealing with numerical aperture, refractive index profile, attenuation and dispersion of signals propagating in the fibre, and (ii) mechanical and geometrical, dealing with strength, shape and form of the fibre.

OPTICAL PROPERTIES

a. *Numerical aperture and refractive index profile*
The numerical aperture (NA) of a fibre is a measure of how much light can be launched into the fibre. It has been shown that NA is given by

$$\sin \theta_a = (n_1^2 - n_2^2)^{1/2} \qquad (10.8)$$

For a graded refractive index fibre (GRIN) the numerical aperture is not a constant but decreases from axis outwards. NA for a multimode fibre typically lies between 0.3 and 0.4, while that for a single mode fibre it is of the order of 0.1. The NA of fibre with parabolic profile is approximately half that of the step index fibre. The refractive index profile gives the functional dependence of refractive index on the radial coordinate. Most common form is a parabolic profile.

b. *Attenuation*

The attenuation in a fibre is due to the following factors:

(*i*) Rayleigh scattering

(*ii*) Absorption

(*iii*) Losses due to microbending, and

(*iv*) Loss at imperfect core-cladding interface

Rayleigh scattering is due to microscopic non-uniformities in the core created by drawing process and varies as $1/\lambda^4$. It imposes the lower limit on the achievable attenuation. Another source of attenuation is the presence of impurities in the core material. Hydroxyl (OH) ions are the dominant absorbers in most fibres. The impurity levels should be typically in the order of a few parts in a billion in good quality fibres.

The remaining two factors are dependent on manufacturing and cabling processes. Microbending loss is due to mechanically induced coupling between modes which are guided in the core to modes in the cladding, which are thereby lost from the core. This loss is enhanced when the spatial period of mechanically induced perturbation along the fibre coincides with the difference in the wave numbers of adjacent modes within the fibre. This difference is of the order of millimeters so that spatially induced coupling may readily occur.

c. *Dispersion*

If a short pulse of light is launched into an optical fibre, it emerges somewhat broader owing to the dispersion along the fibre. The dispersion of a fibre is usually specified in terms of pulse broadening per kilometer of fibre length. There are following causes of dispersion:

Intermodal dispersion:

Intramodal dispersion: material dispersion

wavelength dispersion

Intermodal dispersion is due to different modes having different path lengths, and hence emerge at the end distributed in time. For a step index fibre, the intermodal dispersion can be approximated as

$$\tau_{mod} = (n/c)\triangle/2 \text{ (ns/Km)} \tag{10.9}$$

For a fibre where $n=1.5$, and $\triangle=0.01$, the intermodal dispersion is 2.5 ns/Km.

The intermodal dispersion in GRIN fibre with nearly parabolic profile is given by

$$\tau_{mod} = (n/c) \triangle^2/2 \text{ (ns/Km)} \tag{10.10}$$

which is considerably smaller than that of a step index fibre.

The material dispersion and waveguide dispersion are both parts of chromatic dispersion. The material dispersion arises due to the variation of refractive index of the core on wavelength. The material dispersion is present for all types of fibres. For a source that radiates in a spectral bandwidth $\triangle\lambda$ around λ, the material dispersion τ_{mat} can be expressed as

$$\tau_{mat}=(-\lambda/c)\,(d^2n/d\lambda^2)\,L\triangle\lambda=KL\triangle\lambda \qquad (10.11)$$

where L is the length of the fibre and K has the dimension of ps/(nm. Km). The material dispersion is expressed in picoseconds of pulse dispersion/Km of fibre path per nanometer of source linewidth (ps/nm. Km). For silica based glasses this shows zero around 1.3 μm. Wavelength dispersion is pulse broadening in a fibre due to finite spectral width of the light source, even when the refractive index is assumed to be constant. Wavelength dispersion is very small and can be neglected in a graded index and step index fibres.

The dispersion for a given source/fibre combination is obtained by taking the square root of the sum of squares of the material and intermodal dispersions.

MECHANICAL PROPERTIES

The mechanical strength of optical fibres is intrinsically very high. Optical fibres are usually thin strands, and as such can be bent into very small radius of curvature without significant chance of breakage. The bending radii of less than 10 cm will induce significant bending losses. The bending stresses may induce linear birefringence, which is of particular interest in single mode fibres used in fibre optic gyroscopes.

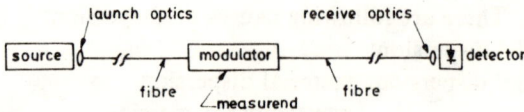

Fig. 10.3 Essential features of an intensity modulation fibre optic sensor.

10.5 A Fibre Optic Sensor Configuration

A configuration of a general fibre optic sensor is given in Fig. 10.3. It consists of a light source which may provide an input beam constant in intensity, frequency, phase, polarization, etc. The light beam is launched into an optical fibre which conducts it to the region of measurement in which the light beam is modulated in one of the aforementioned constant properties. The light is then returned from the measurement zone by another fibre to a detector. The output of the detector is then demodulated. The detectors can not follow the optical frequencies and hence respond to intensity only. Hence any form of modulation is to be converted into intensity modulation before the light is detected.

The fibre optic sensors can be classified into two groups. In the first group fall those sensors in which fibres serve as light conduits to transmit the light to the modulation zone, and to return it to the detector. The modulation process [takes place externally from the fibre, Usually the transmission properties are modulated by the measured parameter (measurand). These fibre sensors are called incoherent fibre optic sensors. In the second group fall those sensors in which the measurand interacts with the light in the fibre. The measurand may modulated phase, polarization and intensity of the light in the fibre. These fibre sensors are called the coherent sensors.

A fibre optic sensor is thus studied in the following parts:

(*i*) Light sources
(*ii*) Light detectors
(*iii*) Methods of modulation
(*iv*) demodulation of light

The important part, that is the fibre itself, has been described in detail in sec. 10.3. The launching of the light in the fibre, coupling between fibres etc. are important procedures and require familiarity. However optical coupling between fibres, and from sources to fibres depends totally on achieving a match between the size and numerical aperture of the core and that of the source. Coupling between fibres of unequal core sizes is usually avoided as it results in significant loss. There are losses due to lateral and angular misalignments of the fibres. The couplers with relatively low loss are available commercially.

10.6 Light Sources

The light sources can be placed under two groups incoherent and coherent sources. Incandescent sources, LED's etc., fall under incoherent sources while the lasers are coherent sources. There are various kinds of lasers available but those used in fibre optic sensors are He-Ne lasers and semiconductor lasers.

There are a few parameters that should be considered while making a choice of a particular source. One of them is the radiance of the source. This is defined as the power radiated per unit area per unit solid angle. High radiance sources are required for fibre optic sensors. Typical launch power in a fibre is one mW. A criterion to determine the amount of power that can be launched into a fibre is 'area multiplied by the solid angle of the fibre and the source radiance'. It may be noted that single mode fibres require the highest radiance of the source. The second parameter of interest is the noise in the source. The minimum noise level is determined by the shot noise which is due to photon nature of light. The noise in the source actually far exceeds the shot noise and there are many reasons for it. The spectral characteristics of the source is another parameter of interest. The influence of environmental parameters on the output and life is also an important parameter for the selection of a source.

The He-Ne laser is most frequently used source. It radiates usually at 633 nm in red; the He-Ne laser can however be operated at a number of wavelengths in the visible region. The output of the laser varies from a few tens of μW to 100 mW range. The laser usually operates in TEM$_{oo}$ mode. It is a very high radiance source. A 1 mW power in a beam of approximately 1 mm diameter and of 1 mrad divergence corresponds to a radiance of approximately 10^8 W/(cm^2-steradian) in a single transverse mode. The laser can easily be operated in a single longitudinal mode.

The semiconductor optical sources are basically of two types viz. light emitting diodes (LED's) and lasers. It is known that *p-n* junctions made from many semiconductor materials from group III and V of the periodic table emit external spontaneous radiation in visible, near and far infra red regions of the spectrum under appropriate conditions. Such devices is are known as LED's. The typical spectral width of such devices is in between 30 to 50 nm. The laser works on the principle of stimulated emission and gives bandwidth in the range of 0.5 to 3 nm.

(a) *LED's as Incoherent emitters*

The LED used in fibre optic sensors are high radiance sources. They are available both homostructure (single material devices) and heterostructure diodes. The heterostructure diodes generate more optical power. LED's structures have been developed which can be classified broadly as

(*i*) surface emitters, and

(*ii*) edge emitters.

Figures 10.4(*a*) and (*b*) show the schematic details of these sources.

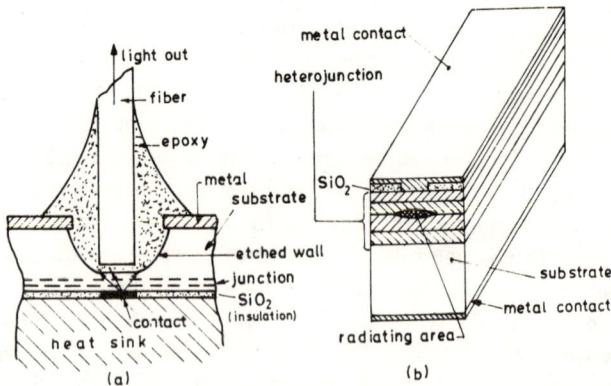

Fig. 10.4 (*a*) Section through the Burrus diode structure.

(*b*) Schematic representation of an edge emitting LED.

The GaAs diffused junction (homostructure) was developed by Burrus et al in 1970, thereafter the same has been implemented at a number of places. The Burrus structure is shown in Fig. 10.4(a). These devices are pigtailed to a fibre, and are excellent sources for multimode optical fibre systems. They are capable of generating about 2 mW of optical power from a surface area of 50 μm diameter in a solid angle of 2π steradian, corresponding to a radiance of 25 W/(cm²·steradian). The coupling efficiency to the fibre is typically 10% due to the generated power occupying too great a solid angle. The radiance of surface emitters is so low that it is impractical to consider them with a monomode fibre systems.

Edge emitters are off shoot of the resonant cavity semiconductor lasers in which the resonant cavity is spoiled giving rise to incoherent radiation. The device operates in a super radiant mode, below the lasing threshold. The output exhibits good spatial coherence. The total power generated by edge emitter is smaller than that from a surface emitter but the radiance is significantly higher. At 1 mW, an edge emitter may have a radiance of over 10^3 W/(cm²·steradian).

The LED may be considered to be a band limited Gaussian noise source, i.e. the radiation is effectively shot noise limited unless spurious noise is transmitted from the bias supply net work. The effect of electrical bias supply on the optical output is of first order: both frequency and output intensity will be function of electrical bias. The LED's may be directly amplitude modulated simply by varying the bias current.

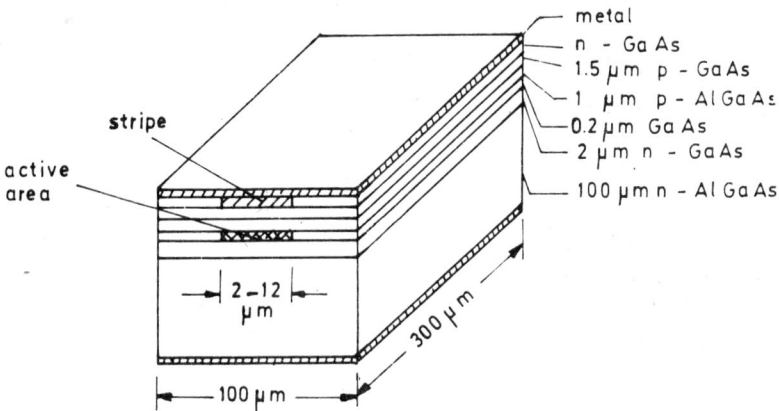

Fig. 10.5 Stripe contact DH laser

(b) *Semiconductor laser*

Semiconductor laser diodes are very versatile sources of optical energy. Fig. 10.5 shows a schematic of a stripe contact DH laser. In a semiconductor laser, the cavity ends provide the feedback due to Fresnel reflection at cavity-air interface; the reflectivity is about 30%. Therefore any reflected beam particularly due to Fresnel reflections from the fibre ends can cause serious fluctuations in the amplitude and phase of the output power. The optical power radiated from laser leaves a rectangular aperture of 0.5 μm in depth and 5 to 25 μm in width of the end of the cavity. The radiation is coherent across this aperture. The field for the fundamental mode is confined to a highly astigmatic core; the half angles cover a range from 30° to 60° vertically and 5° to 20° horizontally. The radiance of these devices is very high. For a laser generating 10 mW from an aperture 10μ m$\times$ 0.5 μm into a cone of 5° by 45° half angles, the radiance is of the order of 10^8W/(cm^2-steradian). Noise in semiconductor lasers arises due to a number of sources; like multimode interference noise, resonance noise, relaxation noise, self pulsing etc.

10.7 Detectors

The detector converts the incident optical energy into the electrical energy. Semiconductor detectors, CCD arrays, photomultiplier tubes (PMT), photoconductors etc., are some of the detectors used with fibre optic systems. Of these semiconductor detectors are small and hence compatible to fibres. Besides they also meet a number of other related requirements namely speed, responsivity, noise and bias voltage.

All photodetectors rely on the energy of incident photon either to create an electron hole pair in semiconductors or to cause the release of an electron from the cathode of a PMT. Due to photon nature of light, the shot noise is inherent in all optical detectors. There is an optical wavelength above which the detector does not respond to the light. This is governed by the equation

$$h\nu \geqslant E_{\text{threshold}} \tag{10.12}$$

The detector may be characterised by its quantum efficiency. This is defined as the fraction of useful electron to the incident photon. A typical value of quantum efficiency is 70%.

Four types of semiconductor detectors are in use in optical systems: the PIN diode, the avalanche photo diode (APD), the PIN-FET hybrid module and photoconductors. The basic detection process is identical in all these devices i.e., the creation of electron-hole pair by incident photons. Therefore the wavelength response of the detector is determined primarily by the band gap of the semiconductor material.

(a) *The PIN diode*

A PIN diode consists of three regions: a *p*-type region, an intrinsic region and an *n*-type region. The free electrons from the *n*-type region diffuse to the intrinsic region. Similarly the holes from *p*-type region

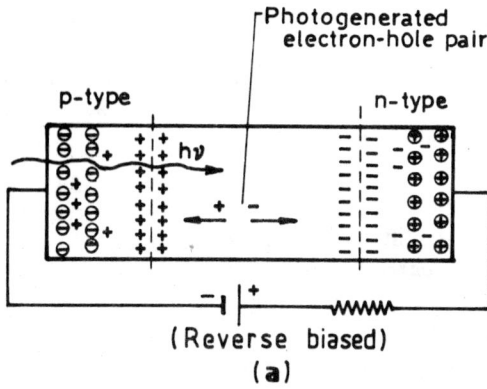

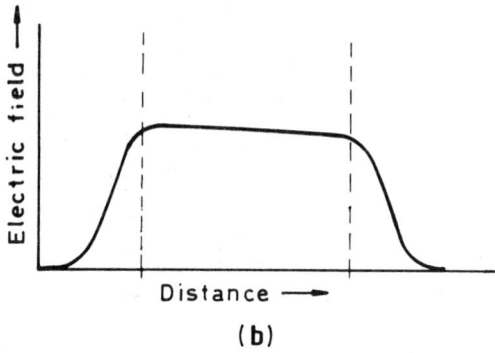

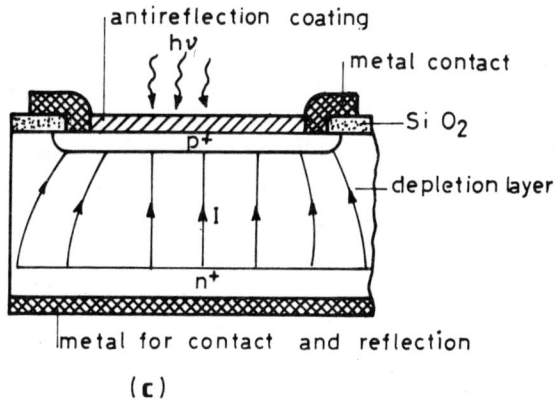

Fig. 10.6 (a) Microscopic details in pin diode.
    (b) Variation of electric field with distance.
    (c) Cross-section of a pin diode.

diffuse to the intrinsic region. This diffusion process, however, does not continue indefinitely. For every free electron leaving *n*-type region, an immobile +charge is left behind in the *n*-type region. The amount of + charge increases as the number of departing free electrons increases. Similarly as the holes depart from the *p*-region, immobile —charges build up in the *p*-type region. Therefore, a potential difference exists between the two regions. This potential is observed externally as contact potential. Fig. 10.6(a) shows the microscopic picture in the three regions, while Fig. 10.6(b) shows the variation of electric field E in the structure. It is this electric field which stops diffusion of free carriers to the intrinsic region. The density of mobile carriers drops rapidly as one moves away from *p-i* and *n-i* boundaries towards the middle of the intrinsic region. That portion of the intrinsic region which has scarcity of mobile charges is called the depletion region. The PIN diode is usually operated with a reverse bias voltage. With this bias an additional electric field is created that reinforces the internal electric field. This field further exerts force on the free carriers, as a result the depletion region expands.

When a photon whose energy Nu exceeds the threshold value, enters the intrinsic region, an electron hole pair is produced. The conductivity of intrinsic region is low, while that of *p* and *n* regions is high due to the abundance of free carriers. When an external potential is applied, the conductivity distribution causes the electric field to build up primarily in the intrinsic region and not in *p* and *n*-type regions. Under the influence of applied field, the photo generated electrons and holes are swiftly swept towards *n* and *p*-type regions respectively, creating a signal current.

Figure 10.6(c) shows a cross-section of a typical PIN diode. The width of the intrinsic region is designed to absorb maximum amount of light at the wavelengh of interest. A convenient width of intrinsic region is from 10 μm to 20 μm for the Si photo diodes having maximum efficiency in 0.8 μm to 0.9 μm wavelength range. The light reaches the absorption region via an anti-reflection (AR) coating and a thin p^+ region of high conductivity. The Si PIN diode is most commonly used photo-detector for other than long haul communication applications.

(b) *AVALANCHE photo detector (APD)*

Fig. 10.7(a) gives the cross-section of an APD photo detector. Its structure is similar to that of PIN diode except for the insertion of an additional *p*-type multiplying region. The intrinsic region is slightly doped to reduce its resistivity to around 300 Ω cm. The *n*-type region is made thinner but is more highly doped than the multiplying *p*-type region. The incident photons are absorbed in the intrinsic region and photo generated electrons and holes drift towards *n*-type and *p*-type regions respectively. It may be noted that the region of highest resistivity

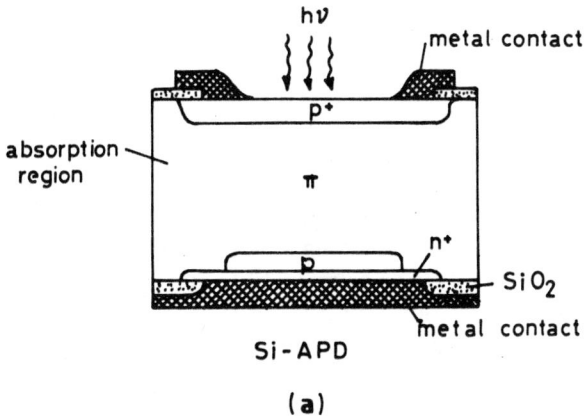

(a)

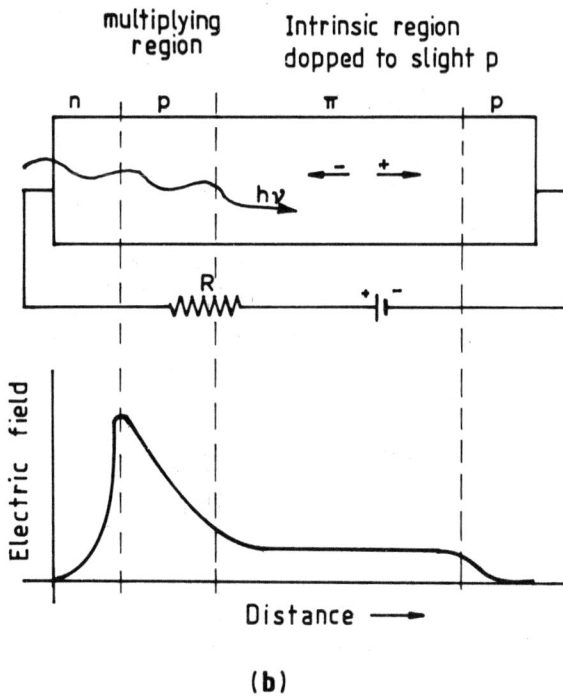

(b)

Fig. 10.7 (a) Cross-section of an APD.
(b) Mechanism of avalanche multiplication and
electric field variation in an APD.

is concentrated near the junction between the *n*-type region and multiplying *p*-type region when reverse bias is applied. The variation of electric field as a function of distance is shown in Fig. 10.7(b). The highest field exists in the region of highest resistivity. The photo generated electrons before reaching *n*-type region, pass through multiplying *p*-type region. If the electrons have acquired sufficient acceleration on reaching the multiplying region, new electron hole pairs are generated by the collision process. The newly created electron hole pairs will collide with the crystal lattice and produce yet further electron hole pairs, thus initiating the process known as avalanche multiplication. Indeed the APD utilizes this process to achieve higher electrical output. A typical gain factor from an APD may be between 10 to 100.

Two major sources of noise in the PIN photo diode and APD are thermally generated noise and shot noise. Thermal noise is independent of signal current whereas the shot noise is dependent on the current. Therefore in practice APD is used in situations where the shot noise, before multiplication is well below the thermal noise. Optimum signal to noise ratio is obtained when the multiplication process brings the shot noise upto the level of thermal noise.

(c) *Photomultiplier tube (PMT)*

The photomultiplier tube is the most sensitive of optical detectors; photon flux as low as one photon per second can be measured. Figure 10.8

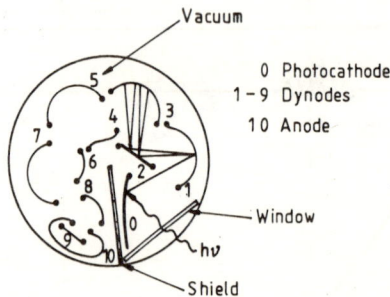

Fig. 10.8 Cross-section of a photomultiplier tube.

shows a cross-section of a multiplier tube. The electrodes in this tube are a photocathode 0, a series of secondary emitters 1 to 9, called dynodes and an anode 10. It is designed for connection to a voltage source and load resistance R as shown in Fig. 10.9. As evident from Fig. 10.9, the dynodes are biased at successively higher potential, usually 100 V per stage. An incident photon whose energy *hv* exceeds the surface work function of the cathode will cause the emission of an electron from the cathode surface. The wavelength response of the tube is thns determined by the material of the cathode. The photo generated electron is accelerated to dynode 1. On reaching dynode 1, each photo generated electron

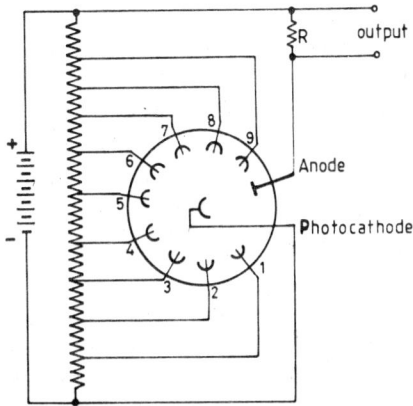

Fig. 10.9 Biasing of various dynodes.

produces N secondary electrons. Attracted by dynode 2 due to its higher potential these secondary electrons produce N^2 new secondary electrons for each original photo-electron. Thus N^9 secondary electrons for each original photo electron reach anode 10. The overall gain of the PMT is thus N^9. In general if there are M dynodes, the gain will be N^M. Typically N is approximately 5 and M between 8 to 10 producing gains from 4×10^5 to 10^7. The gain depends on the bias voltage between successive dynodes. The anode 10 surrounds the dynode 9, a relatively small voltage from dynode 9 to anode 10 is sufficient to collect essentially all the electrons from this dynode. Since photo multiplier tube is a current generator, increasing output resistance R increases output voltage. An upper limit on R may be imposed by the time constant limitation or the non-linearity. The PMT is nearly linear to about 1% for cathode current of 0.1 μA or less. A comparison of various photo detectors is made in Table 10.1.

10.8 Modulation of light

There are a number of methods for modulating the light beams; of these the intensity modulation, phase modulation, frequency modulation and polarization modulation are discussed here.

TABLE 10.1

Device Type	Factors determining detection threshold	Typical operating power range	Wave length range (μm)	Quantum efficiency %	Frequency response
PIN	Photon current exceeds thermal noise current	Usually shot noise limited	0.4 to 1.6 μm depends on material	50	Over 1 GHz available
APD	Multiplied photo current to exceed the thermal noise current	Multiplication of 10 to 100/typical P<100 nW	0.8 to 0.9 μm also available at 1.3 μm and 1.5 μm	50	Over 1 GHz
PMT	Multiplied photon induced electron current must exceed thermal noise current. Gain process is almost noise free	10^{-19} W can be detected. Always use below 1 nW. Higher power can damage cathode	0.3 μm to 1.0 μm	Low typically below 10	100 MHz possible

(a) *Intensity modulation*

Some kind of attenuation of the light is obtained in intensity modulation. This could be achieved externally of the fibre or internally in the fibre. The working of a few methods based on external modulation is described below. Figure 10.10(a) shows an optical sensor using a reflector for external modulation. A part of the reflected energy is coupled to the output fibre. The coupling depends on the position of the reflector apart from fixed parameters like diameters and numerical apertures of the input and output fibres and their separation etc. The intrinsic resolution of this fibre sensor is better than 1 nm. The attenuation can also be introduced by the movement of a mask perpendicular to the direction of the beam. As shown in Figure 10.10(b), the input fibre is imaged onto the output fibre and the mask moves perpendicular to the direction of beam in between the lenses. The mask movement changes the amount of energy coupled to the output fibre. The sensitivity is estimated to be less than 1 part in 10^6. A simpler structure as shown in Figure 10.10(c) operates without the lenses but the sensitivity is reduced due to loss over the gap.

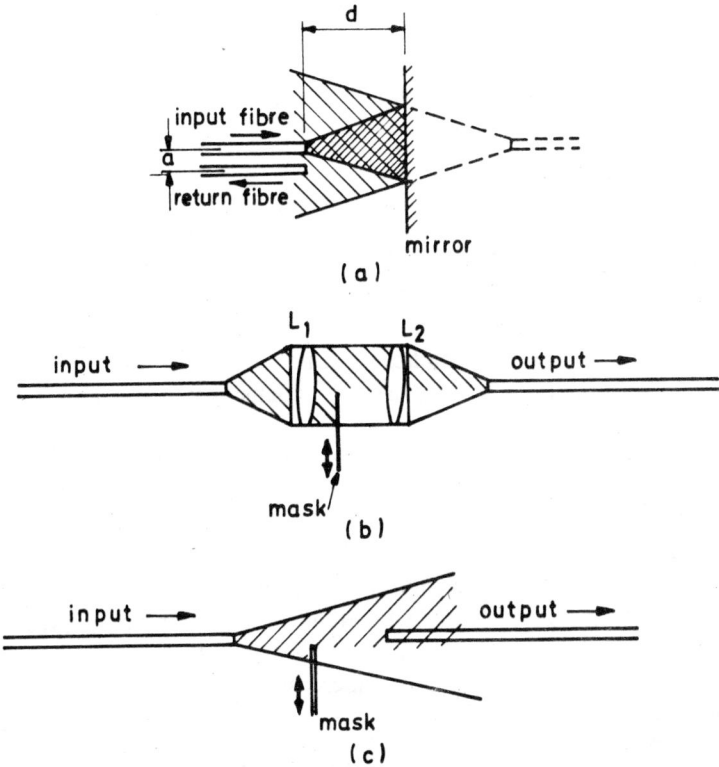

Fig. 10.10 Intensity modulation
(*a*) Attenuation due to reflector position.
(*b*) and (*c*) Attenuation by mask movements.

Simple sensors are based on direct coupling. Figure 10.11 shows schematics of sensors based on position dependent coupling between the tow

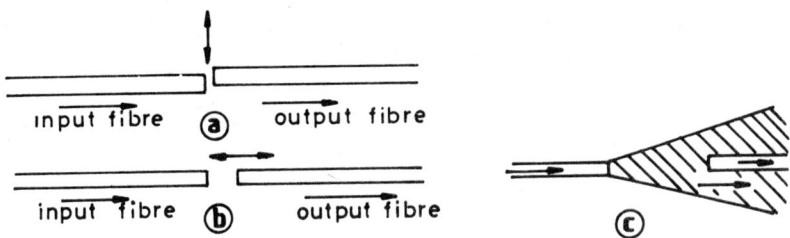

Fig. 10.11 Direct attenuation methods
(*a*) Lateral movement.
(*b*) and (*c*) Longitudinal movement.

fibers which move relative to each other. Both lateral and longitudinal translations of the fibre can be used.

The method described above causes intensity modulation external to the fibre. Internal modulation can be obtained by microbending in optical fibres. It induces losses by coupling the guided modes to the cladding.

(b) *Phase modulation*

The total phase of light along an optical fibre depends on (i) its total length, (ii) the refractive index and index profile, and (iii) the geometric transverse dimensions of the guide.

The phase can be varied by variation of any of these parameters. For example, the total length of the fibre can be modulated by (i) an application of longitudinal strain, (ii) thermal expansion, (iii) an application of hydrostatic pressure causing expansion via the photo elastic effect. The guide dimensions vary with radial strain in a pressure field, longitudinal strain through Poisson's ratio and thermal expansion.

The phase variations are converted into intensity variations using interferometry. The interferometers are used for wide variety of applications. The advantages of using fibres in interferometric sensors lie both in easing the alignment difficulties inherent in assembling long path interferometers, and by increasing the sensitivity of phase modulation to the environmental parameters by simply increasing the optical path exposed to the measured. Use of fibres in interferometry has resulted in compact and mechanically rugged optical interferometers. Fibre optic interferometric sensors are among the most sensitive devices. Hydrophones, magnetometers, accelerometers, strain gauges, and thermometers have been fabricated around fibre interferometer, and all have achieved sensitivities far exceeding than those available from other techniques.

(c) *Frequency modulation*

Frequency modulation occurs under limited range of physical conditions. The principal one of interest is Doppler shift of a beam reflected or scattered from moving targets. Measurement of Doppler shift provides a very sensitive means of measuring velocities. A Doppler probe is capable of detecting velocities over a very wide range from a few μm/sec. to 100 m/sec. It has an extremely large dynamical range and response is linear. As an example, a probe using He-Ne laser will give a Doppler shift of 1.6 MHz per meter per sec. It is frequently used for the measurement of velocity of fluid flows.

(d) *Polarization modulation*

A variety of physical phenomena influence the state of polarization of light. Birefringence may be induced by the application of stress, or field. The phenomena like electro-optic effect, Faraday effect, optical activity etc.. can be used for polarization modulation.

10.8 Demodulation of Light

The optical detectors respond to intensity, that is they are square law detectors. The input optical power is converted to an electrical current in the detection process. Thus the electrical power is proportional to the square of the optical power. Any form of modulation is to be converted to intensity modulation prior to detection. We describe various schemes to detect phase modulation, polarization modulation and frequency modulation.

(a) *Detection of phase modulation*
The detection of optical phase modulation is usually performed with interferometry. The interferometric process converts the phase variations into intensity variations. Consider a Mach-Zehnder interferometer as shown in Fig. 10.12. One arm of the interferometer contains the phase

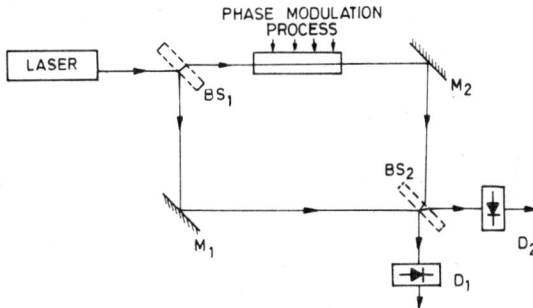

Fig. 10.12 Schematic of a Mach-Zehnder interferometer.

modulation process. On the photo detector D_1, the fields via both the arms of the interferometer add. The resultant amplitude is given by

$$E_t = E_o \sin \omega t + E_o \sin (\omega t + \delta(t)) \qquad (10.12)$$

where E_o is the amplitude of light in each arm at the detector plane, ω is the laser frequency and $\delta(t)$ is the time varying phase introduced by the phase modulation process.

The photodetector responds to the intensity i.e. $(E_t)^2$. Thus the photo current in detector D_1 will be given by

$$i(t) \propto \{1 + \cos \delta(t)\} \qquad (10.13a)$$

The photo current in detector D_2 will be given by

$$i(t) \propto \{1 - \cos \ \delta(t)\} \tag{10.13b}$$

The minus sign occurs because the interference patterns in reflected and transmitted sides are complimentary to each other. We thus find that the phase variation has been converted into intensity variation by the interferometric process, and hence the photo current is a function of phase variation.

The response of the interferometer to small changes in relative phase between the two beams may be obtained by differentiating Eq. 10.13(a). Thus

$$di(t) \propto \sin \ \delta(t) \ d \ \delta \tag{10.14}$$

Thus the change in photo current depends both on the initial phase setting of the interferometer and on the phase change. If we set $\sin \delta = 1$, that is the phases of the beams in two arms of the interferometer are in quadrature, the photo current is directly proportional to phase change. The relationship between the incremental phase change and the incremental intensity change (photo current change) is linear to within 1% over a range of $9°$ at the quadrature condition.

There are, however, two principal difficulties with this interferometric arrangement. Maintaining a stable quadrature bias point is far from trivial, unless the interferometer itself is self-compensated in some way, for instance the Sagnac interferometer. Otherwise some kind of automatic feedback technique is to be used to maintain quadrature condition. The other difficulty is due to the output of the interferometer being sensitive to both the source intensity and relative optical phase. At 2% drift in source intensity produces exactly the same effect as somewhat more than $1°$ phase change. In addition to these difficulties, the interferometers with non-zero path difference require the sources with a very high wavelength stability.

Many of these problems with interferometric detection may be overcome by the use of heterodyne technique. In a heterodyne interferometer, the reference beam is frequently shifted by passing it through a suitable modulator. Figure 10.13 shows a heterodyne interferometer using Bragg cell for imposing a frequency shift ω_B. The phase modulation is induced by vibrating one of the mirrors of the interferometer. The total field incident at the detector is

$$E_t = E_o \sin (\omega - \omega_B) \ t + E_o \sin \{(\omega + \omega_B) \ t + \delta(t)\} \tag{10.15}$$

Therefore the photo current $i(t)$ is given by

$$i(t) \propto \{1 + \cos (2 \ \omega_B \ t + \delta \ (t)\} \tag{10.16}$$

If $\delta \ (t)$ is periodic then it will modulate the intermediate frequency, and standard PM detection techniques may be used to extract the modulation.

Estimates based on the reception of a 1 mW local oscillator signal with a 100 KHz bandwidth indicate a theoretical SNR of about 100 dB. Thus phase modulation in the order of 10^{-5} radians should be detectable. In

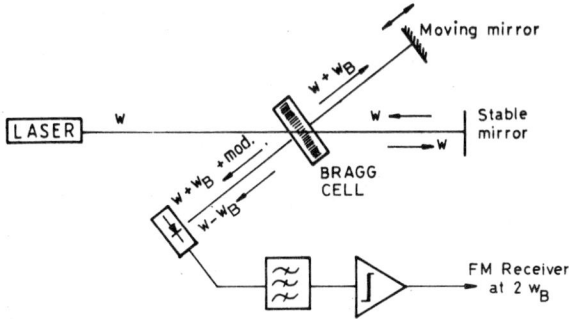

Fig. 10.13 A heterodyne Michelson interferometer.

practice, SNR of 80—90 dB is achievable, and minimum detectable levels in the range of 10^{-4} radians of phase modulation may be obtained.

(b) *Detection of polarization modulation*

The changes in the orientation of lineraly polarized light can be converted into intensity changes by letting it through a polarizer. If the angle between the orientation of the linearly polarized beam and the transmission axis of the polarizer is θ, then the transmitted intensity can be expressed as

$$I_t(\theta) = I_o \cos^2 \theta \qquad (10.17)$$

The incremental change in intensity and the incremental change in orientation $d\theta$ are related through

$$dI_t(\theta) = I_o \sin 2\theta \, d\theta \qquad (10.18)$$

Usually we set $\sin 2\theta = 1$, so that $dI_t(\theta)$ is linearly related to $d\theta$.

In another example, a Wollaston prism is used as a beam splitter. It produces two orthogonally polarized beams which are angularly separated as shown in Fig. 10.14(a). The intensities of these two beams are equal

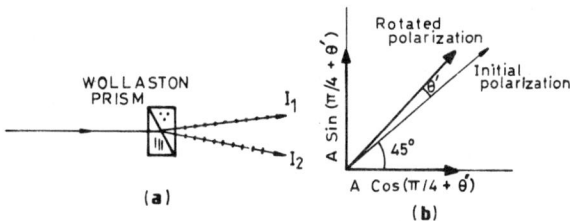

Fig. 10.14 Polarization detection scheme.

when the orientation of the incident linearly polarized beam is at an angle of 45° with the principal section of the Wollaston prism. If the orientation deviates by an angle θ' from 45°, then the amplitudes of the two orthogonal polarized beams are $A \sin (\pi/4 + \theta')$ and $A \cos (\pi/4 + \theta')$ respectively as shown in Fig. 10.10(b). Thus the orientation error θ' is obtained as

$$\sin 2\theta' = (I_1 - I_2)/(I_1 + I_2) \tag{10.19}$$

where I_1 and I_2 are the intensities of these two beams. The measurement is convenient in that it is purely radiometric, and is independent of source intensity fluctuations and variations in the attenuation of the link between the source and the detector. The measurement of a general state of polarization is achieved by the measurement of Stoke's parameters of the incident polarized beam.

(c) *Detection of optical frequency modulation*

It has been mentioned earlier that the frequency of the light scattered from a moving target is Doppler shifted. The frequency shift can be measured using interferometric methods. There are two methods (i) homodyne, and (ii) heterodyne. In homodyning, a reference beam is mixed with the scattered beam. The output carries the Doppler signal. In heterodyning, the reference beam is frequency shifted and then mixed with the scattered beam. The homodyne system has the great advantage of simplicity but cannot yield the sign of the Doppler shift. Whereas the heterodyne system, at the expense of greatly increased complexity, gives complete information on the return signal.

10.10   Some Fibre-Optic Sensors

All fibre optic sensors employ atleast one light source for injecting light into a fibre, a light detector for receiving the signal after light has been modulated by the measurand and demodulated and electronics for converting the detected signal into useful output. The fibre optic sensors can be grouped into two classes: (i) in which fibre merely acts as a conduit to carry signals to and from the sensing area, and (ii) in which the measurand causes the change in the properties of fibre itself. The first case is of external modulation and the sensors are called incoherent sensors. In the second class fall those sensors in which modulation is done internally and these sensors are called the coherent sensors.

INCOHERENT SENSORS

Multimode fibres with LED sources offer simple and reliable sensors for many applications. Some of these sensors are described below:

(a) *Fibre optic position sensor*

A number of variables can be measured using a reflection modulation system (Sec. 10.7a). The variable to be measured is transduced to displacement which is then monitored by fibre-optic sensors. The obvious choice of variables is pressure, temperature, acceleration etc. A sensor known as 'Fotonic' was developed for a number of specialised applications. This is essentially a fibre bundle reflector system in which half the fibres in the bundle are used to illuminate the reflector, and the other half to collect the returned signal. In many situations this sensor may suffer from intensity variations over and above those introduced by the motion of the reflector. This may occur due to tarnishing of reflector, aging of the source, change in connector characteristics, breaking of some fibres in the bundle etc. With the exception of last effect, these problems may be compensated for by a relatively simple arrangement that uses two sets of return fibres separated by a known distance from each other as shown in Fig. 10.15. Measuring the ratio of the returned intensities from each of

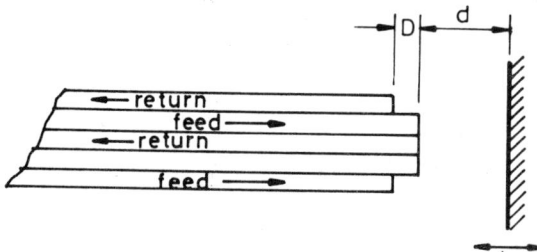

Fig. 10.15 Reflection displacement sensor.

these two separate probes will give unique value for the position of the reflector, provided the correct portion of the sensor characteristics curve is used. In another position sensor, ten output fibres are illuminated by nine input fibres. With appropriate spacing between input and output fibres, resolution of 0.05% of the fibre diameter is possible.

(b) *Optical Microphones*

Numerous schemes have been suggested to realise an optical microphone. In one case, the reflection microphone, the movement of diaphragm modulates the amount of light reflected back along the same fibre. Performance acceptable to telephony has been achieved.

As alternate scheme uses a fixed fibre illuminated by a source and another fibre attached to the core of speaker/diaphragm. The moving core displaces the fibre with respect to the fixed fibre thereby introducing attenuation. This results in varying signal from the detector. For a fibre of 100 μm diameter, a fairly linear response occurs over a movement distance of approximately 10^{-2} cm. The minimum detectable movement is found to be 10 nm and the dynamic range of the microphone is 110 dB.

A more sensitive approach is to place fine absorption grating strips on the opposite ends of the fibres. A transverse displacement equal to grating constant causes 100% modulation.

An alternate scheme based on frustrated total internal reflection (FTIR) has recently been proposed. The schematic is shown in Fig. 10.16, The

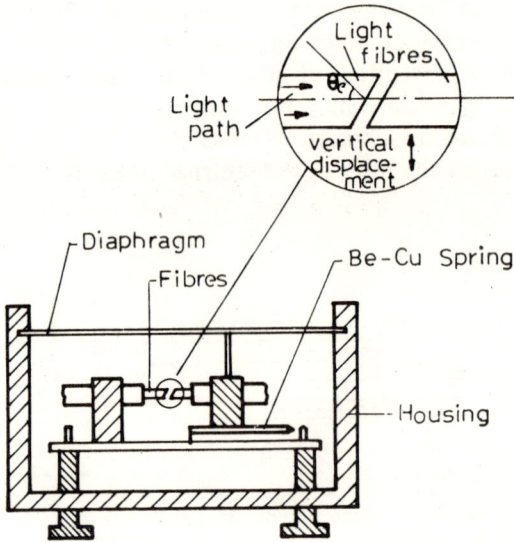

Fig. 10.16 F T IR hydrophone.

ends of fibres are cut at critical angle and polished. The fibres are aligned and placed very close so that appreciable amount of light can be coupled by evanescent wave. The static displacement as small as a fraction of angstrom ean be detected.

(c) *Liquid level sensor*

This has become an established product in the market. It works on the principle that when an unclad fibre is immersed in liquid, the increased refractive index at the interface allows light to escape from the fibre, enabling a go/no-go type detection. These sensors are in market for monitoring levels in petroleum tanks and cryogenic propellent tanks.

(d) *Flowmeters*

This sensor uses a single fibre mounted transversely to the flow within the pipe. The fibre is set into vibration by the natural phenomenon of vortex shedding (Sec. 7.9) causing the phase modulation of light. Using fibres of 200-300 μm diameter, flow rates in the range 0.3—3 meters/sec. have been detected.

COHERENT SENSORS

These sensors measure the optical phase shift introduced by the measurand. The phase modulation is converted to intensity variations using interferometric methods. Fig. 10.17(a) shows a layout of a Mach-Zehnder

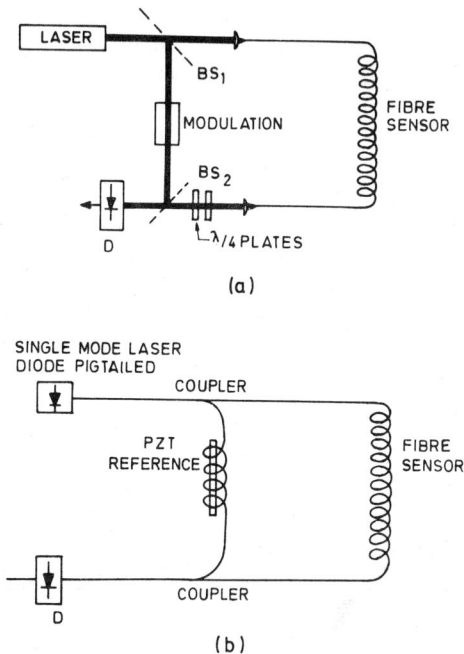

(a)

(b)

Fig. 10.17 (*a*) Schematic of a Mach-Zehnder interferometer having fibre in one arm. (*b*) All fibre Mach-Zehnder interferometer.

interferometer for the measurement of variables that may induce phase variations. The reference beam can be frequency shifted for heterodyning. If good couplers (loss less than 3 dB) are available, a practical sensor is an all-fibre interferometer as shown in Fig. 10.17(b). The advantages of all-fibre interferometers are in mechanical stability, and interface possibility with integrated optics. The limitation on the all-fibre format is in the inconvenience in maintaining the quadrature condition. This has however been successfully implemented by using a PZT cylinder energised via a feedback loop. The heterodyne scheme is almost impossible to incorporate in all-fibre format without breaking the fibre and introducing further mechanically sensitive interfaces.

(a) *Fibre optic Temperature sensor*

There are two properties of the fibre that change with the temperature, namely its dimensions and refractive index. The relationship between the phase change and the temperature change is given by

$$\triangle \delta = - \frac{2 \pi L}{\lambda_0} \left\{ \left(n + \frac{a \lambda_0}{2 \pi} \frac{\partial \beta}{\partial a} \right) \alpha + \frac{\partial n}{\partial T} \right\} \triangle T, \qquad (10.20)$$

where α is the linear coefficient of expansion, L the interaction length, $\partial n/\partial T$ temperature coefficient of refractive index, n mean refractive index of the core, λ_0 wavelength in vacuum, a the core radius and $\partial \beta/\partial a$ the rate of change of propagation constant with core radius. If the glass fibre is used as a sensor, one obtains a phase change of 100 radians per degree Celcius per meter interaction length. This corresponds to roughly 30 fringes per degree Celcius.

(b) *Fibre optic pressure sensor*

An application of pressure differential $\triangle P$ (hydrostatic) to a length L of a fibre changes the phase by an amount $\triangle \delta$ where

$$\triangle \delta = \triangle P \frac{\pi L}{\lambda_0} \left\{ \frac{\lambda_0 a}{\pi} \frac{\partial \beta}{\partial a} - n^3 (P_{11} + P_{12}) \right\} \left(\frac{1 - \mu - 2\mu^2}{E} \right) \qquad (10.21)$$

where P_{11} and P_{12} are photoelastic constants, E Young's modulus and μ Poisson's ratio.

In the interferometer the silica fibre is usually coated with a medium which transforms the pressure field into a longitudinal strain. This constitutes the sensing fibre. This may be in the form of a compliant plastic to enhance the acoustical sensitivity. This coating is drawn into the fibre during manufacture.

The pressure sensor is extremely sensitive. As an example a 10 meter fibre will detect at the threshold of hearing. Further it can be used from DC to 20 MHz. At high frequencies (>100 KHz) the conversion of pressure wave to longitudinal strain modulation decreases in efficiency. The use of coated fibres is therefore restricted to frequencies less than

100 KHz. Between 100 KHz and 10 MHz, the silica fibre may itself be used as a pressure sensor. At still higher frequencies, where the acoustical wavelength becomes comparable to the fibre dimensions, the radial acoustic core mode may be set up. This results in the differential phase term increasing dramatically in multimode fibre, since both compressions and rarefactions are occurring simultaneously in the fibre. Accordingly the differential modulation becomes of the order of twice the peak modulation.

The geometrical flexibility of the fibre hydrophone (pressure sensor) is perhaps the most useful feature. A simple fibre coil will form an omni-directional sensor, provided that the coils diameter is much less than the acoustical wavelength. Two such coils can be assembled, one as a reference arm and the other as a signal sensor arm of the interferometer. The output is proportional to the pressure difference between the two coils. This therefore forms a pressure gradient hydrophone. A single coil may be made highly directional by winding it on a thin long cylinder; the diameter of the cylinder is taken much less than the wavelength and length many wavelengths of the acoustical wave.

(c) *Fibre optic accelerometers*

The fibre optic accelerometers are based on the strain produced in the fibre due to acceleration induced longitudinal stress. The strain causes a phase change which is measured using a Mach-Zehnder type interfero-meter.

In one of the configurations, the accelerometer is in the form of a simple harmonic oscillator consisting of a mass suspended between two fibres or from a single fibre. When the device is accelerated along the length of the fibre, a strain is induced in the supporting fibre due to the acceleration 'a'. For a mass 'm', the force acting is ma. Therefore

$$\frac{\delta L}{L} = \frac{ma}{YA} \qquad (10.22)$$

where $\delta L/L$ is the strain, A cross-section area and Y the Young's modulus of the fibre material. The phase shift $\triangle \delta$ introduced is given by

$$\triangle \delta = \frac{2\pi}{\lambda} \; n \; \delta L = \frac{2\pi}{\lambda} \; n \; \frac{maL}{YA} \qquad (10.23)$$

The sensitivity S of the device is defined as

$$S = \frac{\triangle \delta}{a} = \frac{2\pi}{\lambda} \; n \; \frac{mL}{YA} = \frac{8 \; nm \; L}{\lambda \, Yd^2} \qquad (10.24)$$

where d is the diameter of the fibre. For a typical fibre used in the device, $n=1.5$, $Y=7.3 \times 10^{10}$ N/m², $d=80$ μm, $\lambda=633$ nm, we obtain

$$S = \frac{\triangle \delta}{a} = 4 \times 10^4 \; mL \text{ rad/g} \qquad (10.25)$$

This shows an extremely high sensitivity of the accelerometer.

Fibre optic accelerometer may also be realised using the principle employed in hydrophones. A freely mounted weight may exert pressure proportional to acceleration on a fibre coil. The phase change thus induced may be detected in a Mach-Zehnder interferometer.

(d) *Fibre optic rotation sensor*: *Fibre optic Gyroscope*

Out of all the fibre optic sensors, rotation rate sensor is the most important one having a theoretical sensitivity as high as 10°/hr. For comparison sake, the earth rotation rate is about 14°/hr. Some important features of fibre optic rotation sensor are : (i) it is a passive device, (ii) it has no moving parts, (iii) it consumes less power than other gyros, and (iv) it does not suffer from errors due to null shift and lock-in.

Fibre optic rotation sensor is based on Sagnac effect. It consists of a long single mode fibre in the form of a coil capable of rotating about an axis. Coherent light is launched through both the ends of the fibre as shown in Figure 10.18. The light beams travel in opposite directions and

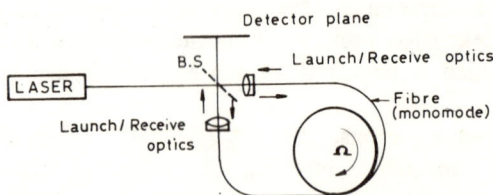

Fig. 10.18 Fibre optic sagnac interferometer.

are collected and made to interfere after one round trip. The fringe pattern in the form of circular fringes is obtained. The fringe pattern remains stationary if the coil is stationary. When the coil rotates about its axis, a phase difference is introduced between the beams travelling clockwise and anticlockwise. The phase difference $\triangle\delta$ is given by

$$\triangle\delta = \left(\frac{4\pi LR}{\lambda c}\right)\Omega \qquad (10.26)$$

where Ω is the angular frequency with which the fibre coil is rotating, L is the fibre length, R the radius of the coil, λ the wavelength and c the velocity of light. The phase difference $\triangle\delta$ causes the fringes to expand or collapse depending on the direction of rotation. However for slow rotation rates, only a small change in the intensity of central fringe is observed. The intensity variation is expressed as

$$I = I_0 \sin \triangle\delta \tag{10.27}$$

where I_0 is the maximum intensity.

The sensitivity of the sensor $dI/d\Omega$ to the variation of Ω can be written as

$$\frac{dI}{d\Omega} = \left(\frac{4\pi LR}{\lambda c} \right) I_0 \cos \triangle\delta \tag{10.28}$$

In Eq. (10.28) the bracketed term can be called the sensitivity of the interferometer in a particular set up. Hence one should use, in principle, long length of fibre in large diameter loops for detecting very small rotation rates. It should not be inferred from this that one can measure any rate of rotation by using longer lengths of fibre in large diameter loops. This is, however, not true in practice since packaging criterion limits the size of the loop, fibre loss sets an upper limit on L, non-linear and damage effects in fibre forbid the use of very high power sources and signal strength, scattered light and quantum efficiency of detector ultimately limits the sensor sensitivity.

It can be shown that for the DC detection scheme, the minimum rotation rate Ω_{min} detectable for a shot noise limited detector is given by

$$\Omega_{min} = \left\{ \frac{4h\,\nu\beta_0}{\eta P_{in}} \right\}^{1/2} \frac{\lambda c}{4\pi\,RL} \exp.\,(\alpha_T\,L/2) \tag{10.29}$$

where h is the Planck's constant, ν the laser frequency, β_0 detector bandwidth, η quantum efficiency of the detector, P_{in} power launched and α_T attenuation of the fibre. If we take $\alpha_T = 2dB$, $L = 4.3$ Km, $P_{in} = 3$ mW, $R = 15$ cm, $\beta_0 = 10$ cm, $\eta = 0.5$ and $\lambda = 633$ nm, we obtain

$$\Omega_{min} = 10^{-4} \text{ deg/hr.} \tag{10.30}$$

Very large changes in the absolute phase of the fibre path are induced by the variations in pressure, temperature etc. The essential feature of the fibre optic gyroscope is that the reference and signal beams in the interferometer travel exactly the same path in the fibre and any air gaps, so that in principle, the phase changes due to environmental variables are fully compensated and the only difference in phase is due to the Sagnac effect. The Sagnac interferometer is a true zero path difference interferometer, so that detection of very small phase difference is fundamentally impossible [unless the operating point of the interferometer is somewhat shifted to the quadrature condition. There are a number of phase detection schemes in use including heterodyne interferometric gyroscope and phase nulling gyroscope.

(e) *Fibre optic Doppler probe*

The frequency of light scattered from the moving target is Doppler shifted. Doppler anemometry has developed into a powerful tool for non-contact measurement of fluid flow velocities over a very large range. It is a contactless and high spatial resolution technique with a linear response. The basic advantage of a fibre optic Doppler probe lies in the fact that the position of the measuring zone may be adjusted without recourse to realignment of the system launch and receive optics. The disadvantage is that the probe may disturb the flow as it is physically inserted, and the probe volume could be a few millimeters in dimensions. Figure 10.19 shows a schematic diagram of a fibre optic Doppler probe. A beam from

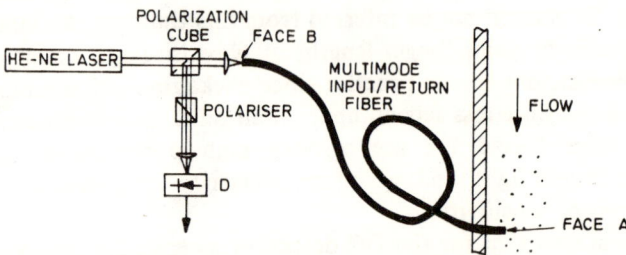

Fig. 10.19 Fibre-optic Doppler anemometer.

a laser source is launched into a multimode fibre via a polarizing beam splitter and launch optics. The other end of the fibre is immersed into a fluid in which the velocity either of the fluid or the bodies within the fluid is to be measured. Light is collected by the fibre and returned. The scattered light is randomly polarized so that half of the returned light is reflected to the detector by the polarizing beam splitter. The reference beam is derived by the Fresnel reflection from the end *A* of the fibre. The multimode fibre very rapidly depolarizes the input light within a distance of a few centimeters. So the returned signal from the end face *A* of the fibre is also depolarized, and hence half of the light returned by reflection from the end face *A* is sent by the polarizing beam splitter to the detector. Thus the reference beam is generated from the correct plane. The two signals, scattered from the fluid and reflected from end face *A*, are photo mixed in the detector whose output is the Doppler signal. The reflection from face *B* of the fibre will be transmitted back to the source by the polarizing beam splitter as this has the same polarization state as the input beam. The fibre optic Doppler probe has been used to measure the blood flow in the veins as well.

(f) *Spectrophone*

This is an instrument used to study the sample of a gas contained in an optical absorption cell. A modulated He-Ne laser beam irradiates the gas, producing a temperature change and hence a pressure change that is ultimately detected. In one actual scheme the pressure fluctuations were detected using a 9.2 meter long single mode fibre wound on a 2.54 cm. diameter tube and placed inside the gas cell. This fibre coil forms one arm of a Mach-Zehnder interferometer. The pressure changes in the cell induced a phase modulation in the light propagating in the fibre coil. A spectrum analyser at the interferometer output provides the analysis of the gas properties.

SENSORS BASED ON POLARIZATION MODULATION

These sensors are based either on the Faraday effect, or the magnetostriction phenomenon. In the latter case, the fibre is coated with a magnetostrictive material during manufacture or a magnetostrictive strip is attached to a silica fibre.

(a) *Magnetic field sensor*
When a linearly polarized light passes through a transparent material of length L which is in the magnetic field H, its orientation is changed. The rotation in the orientation (θ) is given by

$$\theta = V \int_l H_l \, dl \qquad (10.33)$$

where H_l denotes the magnetic field component in the direction of the light beam, dl is the small element of the material and V is the Verdet constant of the material. The measurement of angle of rotation gives the magnitude of the magnetic field. Most suitable materials for this application are dimagnetic materials since in this case the Verdet constant is temperature independent.

The magnetometers using fibres coated with magnetostrictive materials or wound on magnetostrictive mandrel have the potential of far greater sensitivity. Under the influence of magnetic field, the length of the fibre changes which is detected as a phase shift in the output of the interferometer. A change of 10 Gauss in the magnetic field has been detected with a metallic glass jacketted fibre. In another arrangement a single mode fibre is wound under tension on a nickel cylinder which is kept in the magnetic field. The magnetic field, by means of magnetostrictive effect, induces a phase change in the light wave which is detected using Mach-Zehnder type interferometric arrangement. The sensitivity of these devices is very high, approximately 10^{-9} Gauss/meter. Thus one kilometer of fibre length is capable of resolving 10^{-12} Gauss.

(b) *Fibre optic Faraday current monitor*
The current monitor is based on the measurement of rotation of the

linear polarization state due to the magnetic field produced by the current. It is one of the most successful of the fibre optic sensors and is schematically shown in Figure 10.20. The beam from a He-Ne laser is launched

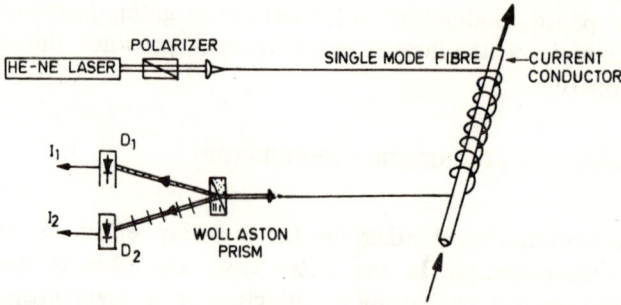

Fig. 10.20 Fibre optic current monitor.

into a single mode fibre. The light is polarized before launch, and in principle, maintains its polarization state until it reaches the current carrying busbar. The polarization is rotated by the magnetic field produced by the current. The beam is then transmitted back to be analysed for the rotation of the linear polarization state. The angle of rotation is related to the current through the magnetic field. The application of this system is on high tension lines, where current and voltage measurements using conventional techniques are both expensive and difficult to implement and interpret.

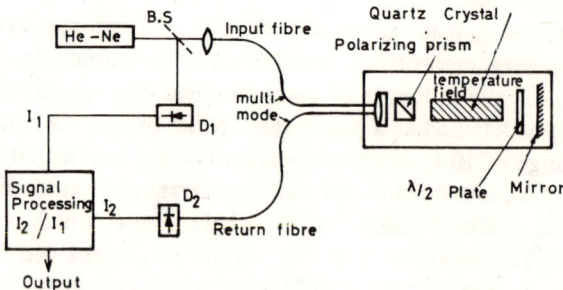

Fig. 10.21 Fibre optic temperature probe.

(c) *Fibre optic temperature sensor*

Figure 10.21 shows the schematic of a temperature probe. A beam from a He-Ne laser is launched into a multimode feed fibre which is at the centre of a bundle of fibres. The light is returned to the detector D_2 via the remaining fibres in the bundle, and the ratio of the intensity (I_2) of the light returned to the light intensity (I_1) launched is calculated. The returned light is modulated by the action of the optical activity of a quartz block which is sensitive to the temperature variations. The laser light has its polarization scrambled on passage through the multimode fibre; the polarization is therefore defined by the polarizing prism. The linear polarization is rotated by an angle θ on passage through the quartz block. The angle θ may be taken as a measure of temperature. The Faraday rotation is also present but its influence is cancelled by double passage through the quartz block. The action of the double passage through the quarter wave plate is to rotate the input polarization. The polarization prism is also used as an analyser. A device having a quartz block of 9 mm diameter and 65 mm length has been fabricated and found to have a resolution of 2°C over a range of 20—180°C. Such devices are again well suited for applications in the electrical power supply industry. It may be noted that the fibres are used in temperature probe only as light conduits.

The fibre optic sensors using polarization modulation can again be grouped under coherent and incoherent sensors.

Exercises

1. The refractive indices of the core and cladding of a fibre are 1.515 and 1.50 respectively. Calculate its numerical aperture and full acceptance angle.
2. A fibre has a numerical aperture of 0.16. Calculate the core diameter of a single mode fibre when used with wavelength of 1.35 μm.
3. Calculate the intermodal dispersion in a step index fibre whose core index is 1.515 and $\Delta = 0.01$. What would be its value in a GRIN fibre if the index 1.515 corresponds to a value on axis?
4. A source of 1 cm² area radiates 100 Watts of power in 2π solid angle. Calculate its radiance.
5. A He-Ne laser delivers 100 mW power in a beam of about 1.4 mm diameter, and divergence of 1 mrad. Calculate the radiance of this laser.
6. The phase of one beam of a Mach—Zehnder interferometer varies linearly with time. Discuss the output of the interferometer.
7. A beam of linearly polarised light is incident on a Wollaston prism. Obtain an expression for the intensity distribution as the prism is rotated. How does it analyse the state of polarization in a current monitor?
8. What is Bragg cell? How does it work? What are the advantages of heterodyne interferometry over the conventional interferometry?
9. Two multimode fibres are in complete alignment. Now one of the fibre is laterally displaced. Using geometric optic consideration, calculate the light coupled as a function of displacement.
10. A fibre of the following specifications is used as an accelerometer:

$$n = 1.5, \qquad Y = 7.4 \times 10^{10} \text{ N/m}^2,$$
$$d = 50 \text{ μm}, \qquad L = 10 \text{ cm}, \quad m = 2 \text{ gms}.$$

Calculate the sensitivity of the device when used with He-Ne laser ($\lambda = 633$ nm).

11. A single mode fibre of 1 km length is wound on a coil of 100 mm diameter. A He-Ne laser beam is launched in the fibre coil. When this coil rotates at the rate of 10^{-3} rad/s, calculate the phase shift in counter propagating beams.

12. A particle is moving with a velocity of 1 m/s. This is illuminated by a beam normal to the velocity vector and the back reflected/scattered light is collected and photo mixed with the direct beam. Calculate the Doppler signal. Assume $\lambda = 633$ nm.

13. A fibre of length of 100 mm is placed in a magnetic field of 1 T. Calculate the angle of rotation of the linearly polarized beam. How would you detect the rotation? Assume $V = 3.17 \times 10^4$.

Reference

CULSHAW B., 'Optical Fibre Sensing and Signal Processing', Peter Peregrinus Ltd., London UK, 1984,

Appendix

Some of the standards* and their accuracies maintained at National Physical Laboratory, New Delhi (India).

S. No.	Parameter (Unit)	Standards/ Technique used	Accuracy	Range
1.	Mass (Kg)	Copy No. 57 of International Proto type kilogram	2×10^{-9}	1 Kg.
		Precision balance	1×10^{-7}	1 Kg.
2.	Length (m)	Kr^{86} discharge lamp	5×10^{-9}	
		Stabilised He-Ne laser	5×10^{-9}	
		Laser Interferometer	5×10^{-8}	1 mm to 1 m
3.	Time/ frequency (sec.)	Portable Cs clock	7×10^{-12}	
4.	Force (N)	Dead weight tester	2×10^{-5}	upto 10^{-5}N
		Lever multiplication of dead weight	1×10^{-4}	upto 1 MN
5.	Pressure (Pa)	Liquid manometer	5×10^{-5}	10^5 Pa
		Micrometer manometer	1×10^{-4}	10 Pa
		McLeod gauge	1×10^{-4}	1 Pa
6.	Temperature (K)			
	Fixed points on International Practical Temperature scale			
6.1		Boiling point of oxygen	0.002 K	90.188 K
6.2		Triple point of water	0.0005K	273.16 K
6.3		Boiling point of water	0.001 K	373.15 K
6.4		Tin point	0.001 K	505.1181 K
6.5		Zinc point	0.001 K	692.73 K
6.6		Silver point	0.1 K	1235.08 K
6.7		Gold point	0.1 K	1337.58 K

*Reproduced by the courtesy of National Physical Laboratory, Hillside Road, New Delhi (India), 1981.

Bibliography

1. Baker H. D., Baker E. A. and Baker N. H., "Temperature measurements in Engineering" John Wiley and Sons Inc. (1961)
2. Backwith T. G. and Buck N. L., "Mechanical Measurements" Addison-Wesley Publishing Co. (1961)
3. Beers Y., "Introduction to the theory of Error" Addison-Wesley Publishing Co. (1957)
4. Benedict R. P., "Fundamentals of Temperature, Pressure and Flow Measurements" John Wiley & Sons Inc. (1969)
5. Cerni R. H. and Foster L. E., "Instrumentation for Engineering Measurement" John Wiley & Sons Inc. (1962)
6. Considine D. M., "Process Instruments and Control Handbook" McGraw-Hill Book Co. (1974)
7. Considine D. M. and Ross, "Handbook of Applied Instrumentation" McGraw-Hill Book Co. (1964)
8. Cook N. H. and Rabinowicz E., "Physical Measurement and Analysis" Addison-Wesley Publishing Co. (1963)
9. Doeblin E. O., "Measurement Systems: Applications and Design" McGraw-Hill Kogakusha Ltd. (1975)
10. Dove R. C. and Adams P. H., "Experimental Stress Analysis and Motion Measurement" Printice-Hall of India Ltd. (1965)
11. Durelli A. J. and Parks V. J., "Moiré Analysis of Strain" Printice-Hall Englewood Cliffs (N. J) (1970)
12. Holman J. P. "Experimental Methods for Engineers" McGraw-Hill Book Co. (1971)
13. Holzbock W. J., "Instruments for Measurement and Control" Van Nostrand Reinhold Co. (1962)
14. Jones E. B., "Instrument Technology" Vol. I Butterworths, London (1965)
15. Kallen H. P., "Handbook of Instrumentation and Control" McGraw-Hill Book Co. (1965)
16. Linford A., "Flow Measurement and Meters" Spon., London (1961)
17. Miesse C. C. and Curth O. E., "Product Engineering, 8, 35 (1961)"
18. Preobrazhensky V., "Measurements and Instruments in Heat Engineering" Vol 1 & II, Mir Publishers Moscow (1980)
19. Tuve G. L. and Domholdt L. C., "Engineering Experimentation" McGraw-Hill Book Co. (1966)

Index